Fishing Grounds

About Island Press

Island Press is the only nonprofit organization in the United States whose principal purpose is the publication of books on environmental issues and natural resource management. We provide solutions-oriented information to professionals, public officials, business and community leaders, and concerned citizens who are shaping responses to environmental problems.

In 2000, Island Press celebrates its sixteenth anniversary as the leading provider of timely and practical books that take a multidisciplinary approach to critical environmental concerns. Our growing list of titles reflects our commitment to bringing the best of an expanding body of literature to the environmental community throughout North America and the world.

Support for Island Press is provided by The Jenifer Altman Foundation, The Bullitt Foundation, The Mary Flagler Cary Charitable Trust, The Nathan Cummings Foundation, The Geraldine R. Dodge Foundation, The Charles Engelhard Foundation, The Ford Foundation, The Vira I. Heinz Endowment, The William and Flora Hewlett Foundation, The W. Alton Jones Foundation, The John D. and Catherine T. MacArthur Foundation, The Andrew W. Mellon Foundation, The Charles Stewart Mott Foundation, The Curtis and Edith Munson Foundation, The National Fish and Wildlife Foundation, The National Science Foundation, The New-Land Foundation, The David and Lucile Packard Foundation, The Pew Charitable Trusts, The Rockefeller Brothers Fund, Rockefeller Financial Services, The Surdna Foundation, The Winslow Foundation, and individual donors.

Special funding for this book was provided by The Bullitt Foundation and The National Fish and Wildlife Foundation.

About The H. John Heinz III Center

The H. John Heinz III Center for Science, Economics and the Environment, a nonprofit institution, continues the work of Senator John Heinz by improving the scientific and economic foundation for environmental policy. The Heinz Center's distinctive contribution is to create new mechanisms for collaboration among the four major sectors essential to solving environmental problems. Industry, environmental organizations, government, and academia all play important roles in defining the agenda that shapes our common future. The Heinz Center provides a venue for these sectors to work together to address challenging issues.

Fishing Grounds

Defining a New Era for American Fisheries Management

Susan Hanna • Heather Blough • Richard Allen
Suzanne Iudicello • Gary Matlock • Bonnie McCay

ISLAND PRESS

Washington, D.C. • Covelo, California

Library of Congress Cataloging-in-Publication Data
Fishing grounds : defining a new era for American fisheries management/
Susan Hanna . . . {et al.}.
      p. cm.
Includes bibliographical references (p.   ) and index.
    ISBN 1-55963-803-6 (cloth : alk. paper) — ISBN 1-55963-804-4 (pbk. : alk. paper)
  1. Fishery management—United States. I. Hanna, Susan.
  SH221 .F59 2000
  333.95′6′0973—dc21

Printed on recycled, acid-free paper ⊚ ✪

Manufactured in the United States of America
10 9 8 7 6 5 4 3 2 1

Contents

Preface

Fisheries management is equal parts economics, politics, and theater.
Richard Young, F/V City of Eureka, *Crescent City, California*

A fishery, in my mind, is the human activity of using the resources. I often hear people say that "fishery management is a people problem," but I don't see evidence of that in the way management is carried out.
Courtland L. Smith, Oregon State University

This book is about U.S. marine fishery management, the transition it faces, and the problems that challenge that transition. These problems are chronic, arising from the history of fishery management, difficulties in maintaining fishery productivity, confusion about ownership, complicated decisionmaking, mismatched incentives, incomplete science, and a failure to evaluate management performance. They all speak to a lingering need for clarification, integration, and resolution.

Fisheries are more than the fish in the sea. They are people interacting with fish and with fish markets, fishing technology, government policy, and marine ecosystems. Fishery management is the attempt to accommodate these interactions in a way that sustains fish populations over time.

Fisheries exist in a state of flux. The dynamics of oceans, variability in fish populations, advances in technology, fluctuations in markets, and evolution of government policy all make change a constant. People in fisheries are familiar with change. But the changes affecting fisheries today are different and larger than those that have come before. They are fundamental changes in the way people think about fisheries and the expectations they have about how fisheries are managed. Many people are unsure about the nature of the problem or the type of change required, and they are uncertain how to adapt. The need for plain talk among fishery stakeholders has never been greater.

In its focus on transition and change in fisheries, this book has application well beyond fisheries. The problems it describes arise in a wide range of interactions between humans and the environment—interactions that have been plagued by uncertainty, shortsightedness, and conflict. People concerned about land use, water allocation, forest management, or other resource issues are likely to find much of value and interest here.

Origin of the Book

This book is a product of the Managing U.S. Marine Fisheries program at The H. John Heinz III Center for Science, Economics and the Environment. The Managing U.S. Marine Fisheries program was established to explore alternative solutions in fishery management and to contribute to the debate surrounding the 2000 reauthorization of the Magnuson-Stevens Fishery Conservation and Management Act. The book, one of several products of the program, was developed to provide an understanding of the background and context of problems facing American fisheries.

In the summer of 1998, the six authors met at The Heinz Center to discuss fishery management and its problems. In the course of that discussion we identified several unresolved issues creating chronic problems in fishery management. We decided to explore the causes of these continuing problems by interviewing a number of people with substantial experience and expertise in U.S. fisheries. During the summer and fall of 1998, we conducted in-depth interviews with 77 people from government, environmental organizations, the fishing industry, and academia, asking them open-ended questions about the issues we had identified.

These interviews with people in fisheries—commercial fishermen, recreational fishermen, seafood processors, environmentalists, government scientists, managers, and academics—form the basis of the book. The people we interviewed represent the full range of fisheries, fishing regions, interests, and experience in fishery management. Some have been active in fishery management since the 1977 implementation of the Fishery Conservation and Management Act, the first comprehensive federal fishery management law. As a group they have many different perspectives on how management should work, as will become clear in the chapters that follow. But it will be equally clear that despite the diversity they are united by a common bond—a strong attachment to fisheries, an appreciation for their potential, and a desire to see them productive. They are all truly stakeholders in the future of American fisheries.

Stakeholder Voices

The voices of stakeholders echo throughout the book in short attributed quotes, in text discussions, and in reflections at the end of each chapter. Their voices are their own—although all are affiliated with particular organizations or interests, statements reflect personal views and do not necessarily represent the positions of their organizations. People were extremely generous in sharing with us their thoughts and perspectives on difficult and controversial management issues.

The stakeholders we interviewed often have widely diverse perspectives. In light of these different, and often strongly held, views, we take particular care to point out that the participation of a person in an interview, or the inclusion of his or her perspective in the book, does not imply support for, or agreement with, all arguments made in the book.

Some of the stakeholder comments may not be supported by the evidence. Nevertheless, we included these comments because they reflect peoples' views and form the basis for their approach to, and perspective on, fishery management. As such, they must be taken into account by policymakers. The reader may agree or disagree with a particular point of view but must acknowledge that it is a view that may demand to be recognized as management decisions are made.

Book Scope and Structure

In the past 22 years of U.S. fishery management, many actions have been taken to control the effects of fishing on fish populations, with varying degrees of success. Overall there have been problems in maintaining the biological health and economic benefits of fishery resources, although the extent of the problem varies widely across regions and fisheries. The scope of this book is the range of those fishery management actions, both past and present.

The book is organized by issues and by time: past, present, and future. We first present an overview of U.S. fishing regions, then a review of the present status of fisheries in those regions, and finally a look at the history of fishery management and its changes over time. We examine how decisions made for different reasons at different points in time shaped the history and evolution of fishery management practices and problems.

We look at the problems of the present. While acknowledging the immediacy of problems such as overfishing, bycatch, and loss of fish habitat, we recognize them as symptoms of deeper, underlying problems. We turn our attention to these causal problems in chapters on fishery productivity, ownership of resources, management decision-making, incentives, scientific information, and evaluation.

We end the book by looking to the future. In the final chapter, we synthesize the lessons learned in the previous chapters into recommendations for facing the challenges ahead. These recommendations relate to making an effective transition from today's fisheries to a more sustainable future, recognizing and promoting the regional diversity and cultural richness that fisheries represent, integrating knowledge and management objectives, learning from successes and failures, and creating expectations for stewardship.

Fishery management is a world of acronyms—ABCs, TACs, FMPs—and it is easy to slip into their use. Acronyms can be confusing and distracting, so we have restricted our use of them to three that represent different versions of the major federal fishery management law. The Fishery Conservation and Management Act (FCMA) was passed in 1976 and implemented in 1977, renamed the Magnuson Fishery Conservation and Management Act (MFCMA) in 1980, and again renamed the Magnuson-Stevens Fishery Conservation and Management Act (MSFCMA) in 1996 with the passage of the Sustainable Fisheries Act. To avoid repeating the full names of these laws we use FCMA, MFCMA, and MSFCMA throughout the text.

Two appendices complete the book. Appendix A lists program participants, including the people we interviewed and the reviewers of this book. Appendix B is an annotated bibliography of selected literature particularly relevant to the subjects we address.

Susan Hanna
Heather Blough
Richard Allen
Suzanne Iudicello
Gary Matlock
Bonnie McCay

Washington, DC
September 1999

Acknowledgments

We gratefully acknowledge the many sources of support for the Background and Context of U.S. Marine Fishery Management project.

The NOAA National Marine Fisheries Service and Surdna Foundation, Inc., provided funding support.

Four senior advisers to the Managing U.S. Marine Fisheries program gave policy guidance, advice, and the benefit of their expertise:

- Captain R. Barry Fisher, president, Midwater Trawlers Cooperative, Newport, Oregon
- D. Douglas Hopkins, Esq., senior attorney, Environmental Defense Fund
- Dr. Andrew A. Rosenberg, deputy assistant administrator for Fisheries, NOAA/National Marine Fisheries Service
- Professor Michael K. Orbach, director, Marine Laboratory, Nicholas School of the Environment, Duke University

Jim McCallum provided helpful liaison and coordination with the National Marine Fisheries Service.

Interview sources freely provided information, insight, and knowledge—the outcome of years of experience in U.S. fisheries and fisheries management.

Manuscript reviewers provided constructive comments and suggestions.

Photographer Peter Ralston [www.pralston.com] generously donated the cover photograph, *Off Matinicus Rock*. The photograph is one of a collection published as *Sightings: A Maine Coast Odyssey*, Down East Books, Camden, Maine, 1997.

Mary Hope Katsouros, senior vice president, The H. John Heinz III Center for Science, Economics and the Environment, provided the idea for the program.

We are indebted to all these contributors and thank them for their gifts to this project. We share with them a dedication to finding a better future for the management of U.S. marine fisheries.

PART I

Introduction
and Overview

We want to be sure that the United States does not lose the opportunity to be a leader in fisheries management, given that we have the ability, resources, and techniques to be that leader.

Andrew A. Rosenberg,
National Marine Fisheries Service

CHAPTER I

U.S. Fisheries Today

The overall status of fisheries nationally and globally is clearly a problem.

Linda Behnken, Alaska Longline Fishermen's Association

U.S. marine fisheries, from New England to the Gulf of Mexico, to the Western Pacific and Alaska, are as varied and diverse as the regions that support them. They reflect the ecosystems, economies, and histories of their regions, the ethnic traditions and culture of the people who fish them, and the social structure of their communities.

The diversity of fisheries is represented in the number of ways fish are caught and used. It is also reflected in regional differences in the status of fish stocks, the condition of fishery economies and fishing communities, and the state of fishery management.

Types of Fisheries

Fish are caught in four general types of fisheries: commercial, recreational, subsistence, and indigenous. The mix of these fisheries varies from region to region. Information is available on the landings and value of commercial fisheries and is estimated for the number of fishing trips and fish caught in recreational fisheries. Little information is available, for the most part only descriptive, on subsistence and indigenous fisheries.

Commercial

Commercial fisheries are those in which catches are sold. U.S. commercial fisheries landed 4.2 million metric tons (a little over 9 billion pounds) of edible and industrial seafood at U.S. ports in 1998, continuing

the downward trend of the past several years. These landings were valued at $3.1 billion ex-vessel—at point of first sale. An additional 182
thousand metric tons (about 401 million pounds), valued at $165.9
million, were landed at ports outside the United States or delivered to
processing vessels at sea. About 78 percent of the fish landed in 1998
were used as fresh and frozen food, with the rest processed into canned
and cured products and nonedible fish meal and oil.[1] These proportions
have been fairly stable over the recent past. The United States continued
as the world's fifth-largest seafood producer in 1998, after China, Peru,
Japan, and Chile, accounting for 4.5 percent of the total world catch.[2]

Commercial fisheries are conducted in every U.S. coastal region by
people using a range of vessel types, vessel sizes, and gear types. Fishing businesses range from owner-operated family vessels to multi-
investor–owned factory trawlers. They also include seafood processors,
brokers, transporters, suppliers, and retailers. Alaska routinely leads all
states in the volume of landings—more than all other states combined—with 4.9 billion pounds sold for $951 million in 1998. Dutch
Harbor–Unalaska, Alaska, was the top producing port in the United
States in 1998, where 597 million pounds of fish were landed, valued at
$110 million.[3]

Recreational

Recreational fisheries are those pursued for pleasure and relaxation.
Recreational catch includes fish that are retained as well as fish that are
released. In 1998 an estimated 17 million marine recreational anglers
made over 60 million trips and caught about 312 million fish.[4] Of these,
136 million fish weighing about 195 million pounds were kept. The
rest of those caught—more than half—were released alive. For the
United States as a whole, recreational landings are small in comparison
to commercial landings—about 2.4 percent—although the ratio of
recreational to commercial catch varies widely by species and by region.
Most recreational anglers, trips, and catches occur on the Atlantic and
Gulf coasts, where recreational fishing is important and continuing to
grow.[5] There are important components of recreational fisheries that
overlap with commercial fisheries in the form of businesses such as
charter and "head" boats that provide recreational fishing for a fee and
businesses such as marine suppliers and hotels that exist to support
recreational fishing.

Subsistence

Subsistence fisheries are those conducted for food and other products
but not for commercial or recreational use. These fisheries are pursued

mostly, but not exclusively, by native peoples. They represent a small proportion of the total catch but are important to particular regions such as Alaska and the Western Pacific for both native and nonnative people.[6]

Indigenous

Indigenous fisheries are established through treaties and other agreements between the federal government, tribal governments, and other native groups. They are concentrated in California, Oregon, Washington, and Alaska. These fisheries include commercial, subsistence, and ceremonial catches. Treaty tribes are formally represented in Pacific fishery management, where the total catch quotas of salmon and several stocks of groundfish and shellfish include direct treaty allocations. Community development quotas in Western Alaska set aside a portion of the total allowable catch of Bering Sea groundfish and crabs for the mostly native villages. Similar programs were authorized for Hawaii and other U.S. Pacific islands in the Sustainable Fisheries Act, which amended the MFCMA in 1996. Indigenous fisheries are a small proportion of total U.S. landings but are of large importance regionally.[7]

Processing, Trade, and Consumption

American seafood processors produced about $7.4 billion of edible and nonedible fishery products in 1998, about 10 percent less than in 1997, continuing the recent downward trend in domestic processed product. As is usual, edible products accounted for most ($6.8 billion) of processed value. In order of volume, the major processed food products are raw fillets and steaks, fish sticks and portions, and breaded shrimp. Also important are canned fishery products, such as tuna, salmon, clams, and sardines.

The United States imported 3.6 billion pounds of edible seafood products in the form of fresh, frozen, canned, cured, caviar, and roe in 1998. These imports were valued at $8.2 billion, an increase of 308 million pounds and $419 million from the previous year. The increase in seafood imports continued the 1990s trend of an expanding trade imbalance in seafood products. Also increased were imports of industrial fishery products such as fish oils, meal, and scrap, valued at $7.4 billion. The United States imports a greater value of food fish from Canada, Thailand, Ecuador, and Mexico than from other countries, the most valuable of these imports being shrimp.[8]

Exports were valued at $8.7 billion in 1998, a decrease from the previous year. Most fishery exports—about 75 percent—are nonedible fish meals and oils. The most valuable food export products were fresh and frozen salmon, herring roe, surimi (minced fish meat), lobsters, flatfish, canned salmon, shrimp, and crabs, in that order. The United States exports the largest value of seafood to Japan, followed by Canada, the European Union, and other Asian countries.[9]

Americans spent an estimated $49.3 billion for 12.0 billion pounds of domestic and imported fish products in 1998 at restaurants, caterers, grocery stores, and industrial suppliers. Of this supply, most (10.5 billion pounds) was edible food and the rest feed and industrial products. The production and marketing of fish products in 1998 is estimated to have contributed $25.4 billion to the U.S. gross national product, continuing an upward trend.[10]

Americans ate an average of 14.9 pounds of seafood per person in 1998, up slightly from the previous year.[11] Most is fresh and frozen finfish, followed by fresh and frozen shellfish and canned and cured product forms. Consumption of seafood in the United States has remained fairly stable over time at 14.5 to 15.0 pounds per person, a lower quantity than is consumed in many other coastal nations.

Fishery Management

Conservation and management of U.S. fishery resources are conducted under the authority of the MSFCMA, Public Law 94-265, as amended. The act has been amended by almost every Congress since its first implementation in 1977 (as the FCMA) to reflect changing management needs and conditions. All fishery resources within the U.S. Exclusive Economic Zone—the ocean area extending to 200 miles offshore from the seaward boundary of each of the coastal states (generally 3 nautical miles, but 9 nautical miles for Texas, Puerto Rico, and the Gulf coast of Florida)—is covered by this act, as are many highly migratory species whose range extends beyond the Exclusive Economic Zone.

Fishing in the Exclusive Economic Zone is managed through regulations developed at state, interstate, federal, and international levels. State, regional, tribal, and federal interests are coordinated in a system of regional fishery management councils, in place since 1977. The management system involves many different fishery interests in decision-making. All coastal states, tribes holding treaty fishing rights, and island territories are represented on the councils in eight fishery management regions: New England, Mid-Atlantic, South Atlantic, Gulf of Mexico, Caribbean, Western Pacific, Pacific, and North Pacific (see figure 1.1).

Under the MSFCMA, the eight regional fishery management councils are required to prepare fishery management plans for the fisheries

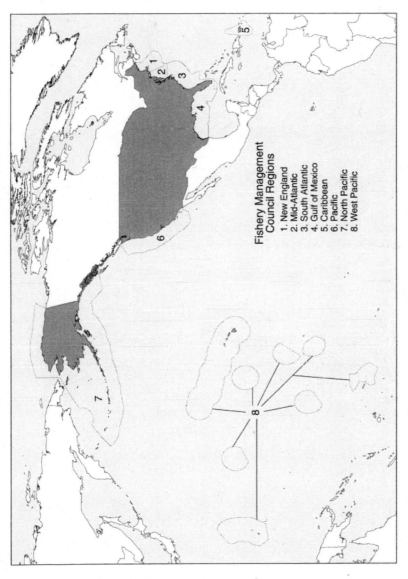

FIGURE 1-1   **New England:** Maine, New Hampshire, Rhode Island, Massachusetts, and Connecticut. **Mid-Atlantic:** New York, New Jersey, Pennsylvania, Delaware, Maryland, Virginia, and North Carolina. **South Atlantic:** North Carolina, South Carolina, Georgia, and the Atlantic coast of Florida. **Gulf:** west coast of Florida, Florida Keys, Alabama, Mississippi, Louisiana, and Texas. **Caribbean:** Puerto Rico and the U.S. Virgin Islands. **Pacific:** California, Oregon, Washington, and Idaho. **North Pacific:** Alaska. **West Pacific:** Hawaii, America Samoa, Guam, and the Northern Mariana Islands. Reproduced based on image provided by the Center for Marine Conservation.

needing management within their jurisdiction. After plans are prepared, they are submitted to the secretary of commerce for approval, which, in most cases, is delegated to the National Marine Fisheries Service. Regulations contained within fishery management plans are enforced through cooperative actions of the U.S. Coast Guard, National Marine Fisheries Service agents, and state authorities. There are currently 39 approved fishery management plans in place and another 7 in development.[12] All fishery management plans and regulations must be consistent with 10 national standards set out in the law and paraphrased here.[13]

Conservation and management measures shall

1. Prevent overfishing and achieve optimum yield (see chapter 3) on a continuing basis.
2. Be based on the best scientific information available.
3. Manage an individual stock as a unit throughout its range and interrelated stocks as a coordinated unit to the extent practicable.
4. Not discriminate between residents of different states. Allocations between fishermen must be fair, promote conservation, and avoid granting excessive shares.
5. Consider efficiency in utilization of fishery resources where practicable.
6. Allow for variations among and contingencies in fisheries, fishery resources, and catches.
7. Minimize costs and avoid unnecessary duplication where practicable.
8. Take into account the importance of fishery resources to fishing communities to provide for their sustained participation and minimize adverse economic impacts on these communities.
9. Minimize bycatch to the extent practicable and minimize the mortality of unavoidable bycatch.
10. Promote safety at sea.

The Eight Fishery Management Regions

New England

The New England Fishery Management Council manages fisheries in federal waters off the coasts of Maine, New Hampshire, Massachusetts, Rhode Island, and Connecticut. The council has 17 voting members—1 from the National Marine Fisheries Service, 5 from state fishery management agencies, and 11 public members appointed by the secretary of commerce. There are 25 stocks under the direct authority of the New England Council. Of these, 11 are overfished, 2 are approaching an over-fished condition, 7 are not overfished, and 5 are of unknown status. These stocks are managed under four fishery management plans:[14,15]

- Atlantic Sea Scallop
- Atlantic Salmon
- Northeast Multispecies
- Atlantic Herring (under review)

The New England Council shares authority with the Mid-Atlantic Fishery Management Council for nine additional stocks. Seven of these (all skates) are of unknown status and are not covered by fishery management plans. The two remaining stocks are managed under joint fishery management plans:

- Monkfish
- Spiny Dogfish

Both of these stocks are overfished. The New England Council has lead management authority over monkfish; the Mid-Atlantic Council over spiny dogfish.[16,17]

The ports of Gloucester and New Bedford, Massachusetts, are the highest producing in the region. In 1998, the greatest volume of fish was landed at Gloucester—107 million pounds, and the highest valued fish at New Bedford—$94 million.[18]

Mid-Atlantic

The Mid-Atlantic Fishery Management Council manages fisheries in federal waters off the coasts of New York, New Jersey, Delaware, Maryland, Virginia, and North Carolina and includes the state of Pennsylvania. The council has 21 voting members—1 from the National Marine Fisheries Service, 7 from state fishery agencies, and 13 public members appointed by the secretary of commerce. In addition to the 9 stocks for which authority is shared with the New England Council, the Mid-Atlantic Council is directly responsible for 11 stocks. Of these, 5 are overfished and 6 are not overfished. These stocks are managed under five fishery management plans:[19]

- Summer Flounder, Scup, and Black Sea Bass
- Atlantic Bluefish
- Atlantic Surf Clam and Ocean Quahog
- Atlantic Mackerel, Squid, and Butterfish
- Tilefish (under development)

The port of Reedville, Virginia, is the highest producing in the Mid-Atlantic region, where 509 million pounds of fish valued at $42.6 million were landed in 1998.[20]

South Atlantic

The South Atlantic Fishery Management Council manages fisheries in federal waters off the coasts of North Carolina, South Carolina, Georgia, and east Florida. The council has 13 voting members—1 from the National Marine Fisheries Service, 4 from the state fishery agencies, and 8 public members appointed by the secretary of commerce. There are 84 stocks under the direct authority of the South Atlantic Council. Of these, 17 are overfished, 11 are not overfished, and 56 are of unknown status. These stocks are managed under five fishery management plans:[21]

- South Atlantic Golden Crab
- South Atlantic Shrimp
- South Atlantic Snapper-Grouper
- Atlantic Coast Red Drum
- Coral, Coral Reefs, and Live/Hard Bottom Habitats of the South Atlantic Region

The South Atlantic Council shares authority with the Gulf of Mexico Fishery Management Council for 10 additional stocks. Of these, 1 is overfished, 5 are not overfished, and 4 are of unknown status. These stocks are jointly managed by the councils under two fishery management plans:[22]

- Gulf of Mexico/South Atlantic Spiny Lobster
- Coastal Migratory Pelagics of the Gulf of Mexico and South Atlantic

The ports of Beaufort, North Carolina, and Key West, Florida, are the highest producing in the South Atlantic region. In 1998, the greatest volume was landed at Beaufort—about 80 million pounds, while landings at Key West had the highest value—almost $45 million.[23]

Gulf of Mexico

The Gulf of Mexico Fishery Management Council manages fisheries in federal waters off the coasts of Texas, Louisiana, Mississippi, Alabama, and west Florida. The council has 17 voting members—1 from the National Marine Fisheries Service, 5 from the state fishery agencies, and 11 public members appointed by the secretary of commerce. In addition to the 10 stocks managed jointly with the South Atlantic Council, the Gulf Council has 61 stocks under its direct authority. Of these, 4 are overfished, 2 are approaching an overfished condition, 6 are not overfished, and 49 are of unknown status.[24] These stocks are managed through five fishery management plans:[25]

- Gulf of Mexico Stone Crab
- Shrimp Fishery of the Gulf of Mexico
- Coral and Coral Reefs of the Gulf of Mexico
- Reef Fish Resources of the Gulf of Mexico
- Gulf of Mexico Red Drum

The port of Empire–Venice, Louisiana, is the highest producing in the Gulf of Mexico region, where 328 million pounds of fish valued at over $38 million were landed in 1998.[26]

Caribbean

The Caribbean Fishery Management Council manages fisheries in federal waters off the coasts of the U.S. Virgin Islands and the Commonwealth of Puerto Rico. The council has 8 voting members—1 from the National Marine Fisheries Service, 3 from the state fishery agencies of the Virgin Islands and Puerto Rico, and 4 public members appointed by the secretary of commerce. There are 179 stocks under the direct authority of the Caribbean Council. Of these, 3 are overfished, 1 is not overfished, and 175 are of unknown status. These stocks are managed under four fishery management plans:[27]

- Spiny Lobster Fishery of Puerto Rico and the U.S. Virgin Islands
- Reef Fish Fishery of Puerto Rico and the U.S. Virgin Islands
- Queen Conch Resources of Puerto Rico and the U.S. Virgin Islands
- Corals and Reef Associated Invertebrates of Puerto Rico and the U.S. Virgin Islands

The municipality of Cabo Rojo, Puerto Rico, is the highest producing port area in the Caribbean region, where about 768 thousand pounds of fish valued at just under $1.6 million were landed in 1998.[28]

Western Pacific

The Western Pacific Fishery Management Council manages fisheries in federal waters off the coasts of the state of Hawaii and the territories of

American Samoa, Guam, and the Northern Mariana Islands. The council has 13 voting members—1 from the National Marine Fisheries Service, 4 from the state and territorial fishery agencies, and 8 public members appointed by the secretary of commerce. There are 57 stocks under the direct authority of the Western Pacific Council. Of these, 3 are overfished, 1 is approaching an overfished condition, 38 are not overfished, and 15 are of unknown status. Most of these stocks are managed under four fishery management plans:[29]

- Western Pacific Crustaceans
- Precious Corals Fishery of the Western Pacific Region
- Bottomfish and Seamount Groundfish of the Western Pacific
- Western Pacific Pelagics

Two stocks of scad are not covered by a fishery management plan.[30]

The port of Honolulu, Hawaii, is the highest producing in this region, where 28 million pounds of fish valued at $49 million were landed in 1998.[31]

Pacific

The Pacific Fishery Management Council manages fisheries in federal waters off the coasts of California, Oregon, and Washington and includes the state of Idaho. The council has 14 voting members—1 from the National Marine Fisheries Service, 1 from an Indian tribe with federally recognized fishing rights, 4 from state fishery agencies, and 8 public members appointed by the secretary of commerce. There are 109 stocks under the direct authority of the Pacific Council. Of these, 12 are overfished, 17 are not overfished, 5 are approaching an overfished condition, and 75 are of unknown status. Of these 109 stocks, 103 are managed under three fishery management plans:[32,33]

- Pacific Coast Salmon
- Pacific Coast Groundfish
- Coastal Pelagic Species

The remaining six stocks are not covered by fishery management plans.[34]

The Pacific Council shares limited management over Pacific halibut with the North Pacific Fishery Management Council. Primary management responsibility for this species is held by the International Pacific Halibut Commission, to which the council makes recommendations about regulations. This stock is not overfished.[35]

The port of Seattle, Washington, is the highest producing in the Pacific region, where an estimated 313 million pounds of fish valued at $22.3 million were landed in 1998. Most of the fish landed in Seattle were caught in the North Pacific. Of those caught in the Pacific region, the greatest volume was landed at Los Angeles, California—145 million pounds, valued at almost $26 million.[36]

North Pacific

The North Pacific Fishery Management Council manages fisheries in federal waters off the coast of Alaska and includes the states of Washington and Oregon. The council has 11 voting members—1 from the National Marine Fisheries Service, 3 from the state fishery agencies, and 7 public members appointed by the secretary of commerce. In addition to the 1 stock managed jointly with the Pacific Council, the North Pacific Council has direct authority for 229 stocks. Of these, 75 are not overfished and 154 are of unknown status. Most of these stocks are managed under five fishery management plans:[37]

- Gulf of Alaska Groundfish
- Alaska High Seas Salmon
- Bering Sea and Aleutian Islands Groundfish
- Bering Sea and Aleutian Islands King and Tanner Crabs
- Alaska Scallops

Two stocks (rattails and sea snails) are not covered by fishery management plans.[38]

The port of Dutch Harbor–Unalaska, Alaska, is the highest producing port in the North Pacific region, where 597 million pounds of fish valued at $110 million were landed in 1998.[39]

Other Fishery Management Authorities

In addition to the eight regional councils, other bodies have responsibility for conservation and management of U.S. fish stocks. Fifty stocks of "highly migratory species" (such as tuna, billfish, and sharks) that fall within the areas of more than one of the New England, Mid-Atlantic, South Atlantic, Gulf, or Caribbean councils are managed by the secretary of commerce through the Office of Sustainable Fisheries of the National Marine Fisheries Service. Of these stocks, 28 are overfished, 20 are not overfished, and 2 are of unknown status.40 These fisheries are managed through two fishery management plans: Atlantic Billfishes and Atlantic Tuna, Swordfish, and Sharks.41

The Atlantic States Marine Fisheries Commission coordinates the management of 21 Atlantic coastal fish species through its Interstate Fisheries Management Program, including lobster, herring, menhaden, bluefish, northern shrimp, red drum, scup, and striped bass. The Gulf States Marine Fisheries Commission coordinates the management of striped bass, Spanish mackerel, blue crab, oyster, black drum, striped mullet, and menhaden.

The Pacific States Marine Fisheries Commission does not have management authority, but it serves as a coordinator of research and as a forum for discussion and liaison between state and federal authorities for issues that fall outside state or regional fishery management council jurisdiction.

The Status of U.S. Fisheries

Overcapitalized, overcrowded, too often overfished; though not currently overfished in the case of the North Pacific.

David Fraser, F/V Muir Milach, Port Townsend, Washington

Overall, I would characterize the situation as "pathetic," given our technical and legal system and all of the intelligence and knowledge we can bring to bear, and the economic potential and the ecological importance of our fish stocks.

Peter Shelley, Conservation Law Foundation

The status of U.S. fisheries is, as with all things, a product of its history. That history has been one of development and growth, stress and decline, taking place at different times in different fisheries. Regional choices and national policies have influenced how, why, and when fisheries have developed and how well they have been sustained. The historical paths have included foreign dominance, domestic territorial claims, and eventual domestic control. They have also included wide-scale regional variation in approaches to fishery management.

How are U.S. fisheries faring? Are they "in pretty sorry shape" or "both good and ugly," as people we interviewed see them? The answer depends on the region in question and whether it is the fishery's biology, socioeconomics, or management that is evaluated.

How stakeholders see the biological status of fisheries is influenced by the region in which they live. Some regions have severe stock depletions. In other regions the stock status is healthy or mixed, but perceived to be below where it should be. And in still other areas, stock status is largely unknown due to a lack of adequate funding for assessment. But, on the positive end, some stakeholders observe that, although there are lots of problem areas, stocks have been amazingly robust. According to one stakeholder, "There are important bright spots where stocks appear to be healthy and stable, and there is a perceptible upward trend in rebuilding depleted stocks. Every time managers have gotten more conservative and effort has declined, we get more fish."

The biological status of U.S. fisheries varies widely. Nationwide, of those stocks for which status is known, 30 percent are overfished, 3 percent are approaching an overfished condition, and 67 percent are not overfished.[42] Important regional differences underlie these numbers; for example, none of the 229 stocks in the North Pacific are overfished, in contrast to New England, where 11 out of the 25 managed fish stocks are overfished.[43] Another way of expressing the biological condition of fisheries is to look at recent yields in comparison to their long-term potential—for the nation as a whole yields are about 37 percent below this potential.[44] The biological condition of many U.S. fish stocks—65 percent—is unknown, despite the fact that most are included in fishery management plans.[45] Fortunately, although large in number, these stocks are small in proportion to total U.S. landings, estimated at less than 3 percent between 1995 and 1997.[46]

While the biological status of fisheries is mixed, the socioeconomic status is not. U.S. commercial fisheries are almost universally overcapitalized—with too much fishing power—and suffering from losses in potential productivity.[47,48] There are more vessels, people, and days spent fishing than are needed to catch the available fish. This means that fish cost consumers more than necessary, and fishermen earn less from the catch than they could if fleets were more efficient. Management also costs more than necessary, because time and resources are spent trying to develop detailed regulations to accommodate overcapitalized fleets and contain the effects of their fishing. Poor economic conditions in a fishery also lead to social problems like conflicts over allocations, loss of jobs, and erosion of traditional culture. As one observer sees it, the poor conditions aren't necessarily leading to positive action: "Leaders continue to spend most of their time in reactive in-fighting on allocation issues, ignoring the real imperative of trying to restore biological productivity and jobs."

This pessimism about the socioeconomic state of fisheries is remarkably consistent across regions. Fisheries are seen as being in poor economic condition even where stocks are biologically healthy because of the extent of overcapitalization and uncertainty about the future. Various problems are affecting the economic condition of fisheries, including declining stocks, growth overfishing (catching fish at too small a size), recruitment overfishing (catching more fish than can be replaced as spawners), bycatch (unintended catch of small or nontarget species), reduced profits, and continuing conflicts over who should be allowed to catch the fish. These are problems that require immediate management attention and often keep managers focused on short-term needs at the expense of long-term objectives. Some of the people we interviewed are concerned that these conditions have left fisheries destabilized, spiraling into more destructive patterns of use. "I don't remember a time when the internal social and business structure of so many fisheries has been so destabilized," says one fishery manager. Many worry about the pressures on fisheries and are uncertain as to whether fisheries will be able to recover or will continue their downward trend.

The biological condition of a fish stock and the economic condition of the fishery that uses it are interdependent. The biological characteristics of a stock affect the economics of fishing because they influence how many fish can be caught, retained, processed, sold, and eaten. These, in turn, affect the number of jobs created, supplies purchased, and money invested. The economics of fishing affects the biology of fish because it drives the behavior of fishermen, processors, and managers, which eventually has impacts on fish populations. Both biology and economics are linked to social factors through their influence on the social well-being of fishing communities. Fishery management ties all of these together and affects the biological, economic, and social status of fisheries through the actions it takes.

In recognition of these interactions, the National Research Council report *Sustaining Marine Fisheries*[49] recommends a stronger integration of the biological and human fishery components through management changes that include reducing fishing mortality and capacity, redesigning institutions, and moving toward a precautionary ecosystem-based approach to rebuilding economically healthy fisheries for the long term.

The status of fishery management is seen by almost all the stakeholders we interviewed as weak. "The whole fishery management system is tragic right now," says one observer. "It's a huge institution, headed in its own direction. It's hard to change." Stakeholders from all sectors are concerned that management seems unable to maintain large stock sizes, in part due to uncertainty about fish stocks and the predisposition to take risks when information is incomplete. A fishery manager worries "about the process of progressively moving down the food chain," when stocks are sequentially overfished. But a few see managers themselves as part of the problem and in denial of reality. "Victories are being proclaimed in an overall context of continued management failure," says an environmental representative.

Very few people we interviewed find reason to defend the status quo of fishery management, and many find it distressing, given the ecological and economic importance of fish stocks to the nation. Stakeholders believe that the poor state of management contributes to weak resources, diminished flexibility, and a loss of economic vitality to coastal communities. Many think that although we are beginning to understand the need to manage fishery resources as part of an ecosystem, we lack adequate information to understand all the biological, economic, and ecological interactions that ecosystem-based management implies. "Given all the money that has been invested in trying to understand and manage fisheries, we've done a pretty poor job," concludes one researcher.

A number of recent publications produced by environmental organizations also focus attention on the failure of management and the number of overfished stocks.[50] Together they focus on the management choices and economic conditions that created incentives to overfish, led to overcapitalized fleets, increased bycatch and discards, and created pressures to take risks with fishery resources.[51]

Conclusion

> Overall the mood is pretty low, a sense of crisis, compared to past years when people felt good about many fisheries.
>
> *Christopher M. Dewees, University of California, Davis*

The picture of U.S. fisheries that emerges from the statistics, interviews, and our own observations is one of fisheries engaged in a difficult transition from a past of development and expansion, with single-species management the practice, to a future that entails stabilization, contraction, and ecosystem-based management practices. The diversity of fishing regions, the pressures on resources, the variability of management performance, and the varying status of fish stocks all add to the challenge of making the transition.

Notes

1. NMFS. 1999. *Fisheries of the United States, 1998.* Current Fishery Statistics No. 9800, U.S. Department of Commerce, National Oceanic and Atmospheric Administration.
2. Ibid.
3. NOAA. 1999. Dutch Harbor–Unalaska Is Nation's Top Fishing Port for 1998. Press release NOAA 99-051, July 14, 1999.
4. NMFS. 1999, n. 1.
5. Ibid.
6. NRC. 1999. *Sustaining Marine Fisheries.* National Academy Press, Washington, DC.
7. Ibid.
8. NMFS 1999, n. 1.
9. Ibid.

10. Ibid.
11. NOAA. 1999. Americans Ate More Seafood in 1998. Press release NOAA 99-49, July 12, 1999.
12. NMFS. 1998. *Report to Congress: Status of Fisheries of the United States.* U.S. Department of Commerce, NOAA, October.
13. NMFS. 1996. *Magnuson-Stevens Fishery Conservation and Management Act.* NOAA Technical Memorandum NMFS-F/SPO-23, December.
14. NMFS. 1998, n. 12.
15. New England Fishery Management Council at http:www.nefmc.org.
16. NMFS. 1998, n. 12.
17. Dan Furlong, executive director, Mid-Atlantic Fishery Management Council. Personal communication.
18. NMFS. 1999, n. 1.
19. NMFS. 1998, n. 12.
20. NMFS. 1999, n. 1.
21. NMFS. 1998, n. 12.
22. Ibid.
23. NMFS. 1999, n. 1.
24. NMFS. 1998, n. 12.
25. Ibid.
26. NMFS. 1999, n. 1.
27. NMFS. 1998, n. 12.
28. Daniel Matos, Puerto Rico/NMFS Cooperative Fisheries Statistics Program. Personal communication.
29. NMFS. 1998, n. 12.
30. Ibid.
31. NMFS. 1999, n. 1.
32. NMFS. 1998, n. 12.
33. Pacific Fishery Management Council at http://www.pcouncil.org.
34. NMFS. 1998, n. 12.
35. Ibid.
36. NMFS. 1999, n. 1.
37. NMFS. 1998, n. 12.
38. Ibid.
39. NOAA. 1999, n. 3.
40. NMFS. 1998, n. 12.
41. NMFS. 1999. Amendment 1 to the Atlantic Billfish Fishers Management Plan. (April) HMS Division, Office of Sustainable Fisheries, NMFS, Silver Spring, MD; and NMFS. 1999. Final Fishery Management Plan for Atlantic Tuna, Swordfish, and Sharks. (April) HMS Division, Office of Sustainable Fisheries, NMFS, Silver Springs, MD.
42. NMFS. 1988, n. 12.
43. Ibid.
44. NMFS. 1996. *Our Living Oceans. Report on the Status of U.S. Living Marine Resources, 1995.* U.S. Department of Commerce, NOAA Technical Memorandum. NMFS-F/SPO-19, February.

45. NMFS. 1998, n. 12.
46. NMFS. 1999. *Our Living Oceans. Report on the Status of U.S. Living Marine Resources, 1999*. U.S. Department of Commerce, NOAA Technical Memorandum NMFS-F/SPO-41, June.
47. NMFS. 1996, n. 44.
48. NRC. 1999. *Sharing the Fish: Toward a National Policy on Individual Fishing Quotas*. National Academy Press, Washington, DC.
49. NRC. 1999, n. 6.
50. For example: Safina, C. 1997. *Song for a Blue Ocean*. Henry Holt, New York; McGuinn, A. P. 1998. *Rocking the Boat: Conserving Fisheries and Protecting Jobs*. Worldwatch Paper 142. Worldwatch Institute, Washington, DC.; Fordham, S. V. 1996. *New England Groundfish: From Glory to Grief*. Center for Marine Conservation, Washington, DC.
51. Iudicello, S., M. Weber, and R. Wieland. 1999. *Fish, Markets and Fishermen: The Economics of Overfishing*. Island Press, Washington, DC.

PART II

The Past

If you don't know where you want to go, how are you going to get there? . . . We don't have a common under-standing of where we are, how we got here, and how we'll get out of the box.

Richard Young, F/V *City of Eureka*,
Crescent City, California

CHAPTER 2

History of Federal Fishery Management

> The U.S. has been the world leader in the past on many environmental issues, but we have a long way to go with fisheries.
>
> *Gerald Leape, Greenpeace*

How did U.S. fisheries get where they are today? Their histories reflect the economic and political histories of their regions. Fishing takes place within the context of larger social and economic events, so what happens in fisheries is in many ways determined by what happens in regional economies, U.S. policy, and international relations. Choices made at these different political levels have either directly or indirectly determined the development path that U.S. fisheries have followed. That path has resulted in many successes, but it has also created problems.

Changes in fisheries have also been driven by fishery conditions, special interests, and changing public values. Over time, the way people look at fisheries has broadened from fisheries as food production systems to fisheries as part of marine ecosystems. Accompanying these broadening perceptions have been expanded expectations for fishery management.

The post–World War II fisheries environment has been particularly dynamic, with changes in technology, markets, information processing, government policies, and public values. This period can be roughly divided into four shorter periods: the 1950–60s, before the FCMA; the 1970s, the decade of the FCMA; the 1980s, the end of expansion; and the 1990s, a period of contraction and change.

1950s and 1960s: Pre-FCMA Decades

> The biggest problem we're facing is that the marine fisheries have
> been the equivalent of the cold war. We built up an enormous fleet
> with societal encouragement, and now we're faced with this enor-
> mous task of building down. It's like deciding what to do with
> these warheads.
>
> *Michael L. Weber, freelance writer and researcher*

American commercial fisheries, in concert with the rest of the economy,
emerged from World War II in a climate of great optimism. Wartime
technological advances like synthetic fibers for nets and fish-finding
electronics made fishing cheaper, safer, and more effective. Fishing vessels that had been requisitioned for war use were reconverted to fishing,
and high wartime fish prices carried over for a short time. But this
optimism faded by the recession of the 1950s, replaced by concern over
low prices and competition from foreign markets. The U.S. commercial
fishing fleet was old and outmoded, unable to compete successfully
with fleets of other nations and the subsidies of their governments.
Pressures built for government policies to rehabilitate the fleets.

Fishing was controlled through state and interstate management in
nearshore waters, and through international agreements outside the U.S.
territorial sea (then 3 miles throughout most of the country). The 1945
Truman Proclamation had unilaterally declared U.S. jurisdiction over the
continental shelf and created a Fisheries Conservation Zone—a contiguous zone of unspecified extent. This proclamation was made in response
to the longstanding problem of Japanese salmon fishing outside the U.S.
territorial sea off Alaska. It was never implemented but did serve as an
encouragement for negotiations and set the stage for extended jurisdictional claims to follow. A second proclamation asserted jurisdiction and
control over continental shelf mineral resources.[1]

By the 1960s the need to revitalize fleets had been augmented by
concern over increasing seafood imports and the desire to protect U.S.
fishing interests from high levels of foreign fishing off the east and west
coasts. Foreign catch off American shores was exceeding domestic
catch,[2] and in the decade from 1958 to 1968, imports had increased
from 37 to 76 percent of the total U.S. seafood supply.[3] Control over
foreign fleets, implemented through international arrangements such
as the International Commission for the Northwest Atlantic Fisheries
and the International North Pacific Fisheries Commission, was considered inadequate to protect U.S. interests.[4,5] By 1966 the United States
had established a fisheries zone of 9 miles contiguous to the territorial
sea, for a total of 12 miles extending out from shore.[6]

The federal government also directed attention to helping the fishing industry expand. Senator Warren Magnuson of Washington, who would later become a principal sponsor of the FCMA of 1976, addressed a 1968 conference on the future of the U.S. fishing industry: "You have no time to form study committees. You have no time for biologically researching the animal. . . . Your time must be devoted to determining how we can get out and catch fish. Every activity . . . whether by the federal or state governments, should be primarily programmed to that goal. Let us not study our resources to death, let us harvest them. . . . To the scientist . . . do they [research products] contribute to the hoped-for American harvest increase? To the agencies . . . are the management regulations you now promulgate designed to help or hinder the American fishermen?"[7]

In 1969 the Commission on Marine Science, Engineering, and Resources—the "Stratton Commission"—reinforced this view when it performed the first systematic review of U.S. policy for the coasts and oceans. Its report on living resources emphasized the national interest in fishery rehabilitation, development, and expansion to create national wealth and meet the world demand for food. It also noted the need to improve stock assessments and better coordinate management authority between the states and federal government.[8]

In the context of the Cold War, strengthening the competitiveness of American industry and control over ocean territory were both important. The ocean was seen as the "new frontier" of resource development. The recommendations to develop and expand the American use of fishery resources that were to follow in the 1970s fit smoothly into this context.

1970s: The Fishery Conservation and Management Act

> We were encouraged by the government to catch more fish. They threw the foreign fishing fleets out so it was up to us to catch the fish.
>
> *Jim Kendall, New Bedford Seafood Coalition*

> The [Fishery Conservation and Management Act] was political and it waffled on who had responsibility and authority for management.
>
> *Lee J. Weddig, formerly National Fisheries Institute*

One of the Stratton Commission's recommendations was to form a single independent agency to coordinate all marine activities. In 1970, President Nixon gave partial form to this recommendation in the creation of the National Oceanic and Atmospheric Administration within the Department of Commerce. What had been the Bureau of Commercial Fisheries in the Department of the Interior became the National Marine Fisheries Service under the newly formed National Oceanic and Atmospheric Administration.

Concerns about the competitive status of American fisheries and the control over ocean territory remained. With the 1973 Eastland Resolution, Congress committed to provide "all support necessary" to strengthen the fishing industry.[9] The third United Nations Convention on the Law of the Sea began a multiyear debate over problems with distant water fisheries off coastal states and the need to implement national fishing zones extending 200 miles from shore.

By 1976, the Law of the Sea Convention had failed to resolve important points of difference between nations. This failure, combined with the highly visible and alarming presence of large foreign fleets off the northeast and northwest coasts and their subsequent depletion of fish stocks, created political pressure for the United States to take unilateral action. State regulation of U.S. fisheries was inadequate to control fishing effects. Senator Magnuson described existing bilateral agreements as "resembling a crazy patchwork quilt of pieced together remnants."[10]

Congress moved to protect domestic fishing interests by enacting the most significant fishery management change in the postwar period—the FCMA. Stating that "the fish off the coasts of the United States constitute valuable resources that contribute to the food supply, economy, and health of the nation," the act established a Fishery Conservation Zone that extended the limit of U.S. fisheries control from 12 to 200 miles offshore.[11] This extension of fishery jurisdiction was the first taken by a major power, to be followed soon after by many other fishing nations.[12]

In addition to expanding the area of federal authority over fisheries, the FCMA established a new structure of eight regional fishery management councils to exercise this authority (see chapter 1). The council system was designed as a decentralized, democratic structure within which fishery user groups and the public would participate as advisers and decisionmakers, and the councils were to become the platform for the major events and decisions that were to follow. The Eastland Commission, charged with surveying the needs of the fishing industry and fishery management in all regions, reflected the sentiment for economic recovery of the U.S. industry in its recommendation for a focus on food production, economic assistance, and federal funding for fishery management. But it also identified emerging environmental and management problems in its recommendation that attention be paid to the decline in fish habitat, council representation, and federal funding for research and management.[13]

The extension of U.S. fishery authority to 200 miles was a major step in reducing the international open access problem, but it created another problem for which the management system was ill-equipped—the need to allocate fish among competing American interests. In retrospect, "the act left too much to the discretion of the councils," one stakeholder observes. The pre-1976 focus had been on getting the foreigners out, aided by the imagery of small-scale traditional fishermen pitted against large-scale foreign fleets.[14] The emphasis of the FCMA on "Americanizing" the fisheries created the impression that its primary purpose was to redistribute fishery resources from foreign to domestic fleets and to reduce the reliance on imported fish through expansion of domestic catches. The council system wasn't prepared to cope with competition among Americans.

In attempts to provide opportunities to all who wanted to fish, some councils put development and expansion before conservation. Even in areas where pre-FCMA levels of foreign fishing had left stocks in need of rebuilding and incapable of absorbing increased fishing effort, federal programs continued to assist industry renovation and expansion. These included vessel construction loans, tax-deferred vessel repair and construction programs, fuel tax relief, gear replacement funds, gear research and development, market expansion programs, and technical assistance programs.[15]

As many people note, U.S. fisheries have been a success story because they were intended to develop and grow. But to support continued growth, government and industry invested in short-term results rather than long-term strategies, and some managers made the mistake of thinking they had enough information to manage quite close to the edge—to squeeze out all possible surplus. Some we interviewed claim that the political power of the fishing industry, disproportionate to its relative size, ensured this management outcome. Several note the capacity-expanding effect of subsidies provided through the Capital Construction Fund[16] and the Fishing Vessel Obligation Guarantee Program.[17] The high levels of fishing capacity that resulted led to pressures to keep catch levels high and access to the resource open.

The "Americanization" of the fisheries—in which foreign vessels were removed in stages from fishing—is seen by some we interviewed as a major contributor to the development of the U.S. fleet. But it happened without a clear idea of where fisheries were going or how fisheries tie into our national interest. "Americanization never really made any sense," says one researcher. "It's an emotional thing."

Many of those we interviewed look back to the early days of FCMA implementation as the source of a host of contemporary management problems. They point to poorly resolved areas of authority and responsibility, describing how from the beginning there has been a lack of clear division of authority between the regional councils and the secretary of commerce. The answer to Who's in charge? was never resolved, to the detriment of effective council functioning. Others note the continuing problem of interference in council actions through congressional or constituent pressure—a problem that affected the ability of the National Marine Fisheries Service to invoke its authority to reject inadequate plans. People with long-time management experience also point to the early and continuing failure to fund the management system at a level commensurate with its expanding responsibilities.

1980s: The End of Expansion

> We have treated fishery management like we treat credit cards: overspend, overspend, pay on the interest, and sooner or later we are bankrupt.
>
> *Roger McManus, Center for Marine Conservation*

By the 1980s the rate of fishery expansion was slowing in most regions. Even though many fisheries were showing signs of strain, continuing concerns over the lack of competitiveness of the fishing industry led to the passage of the 1980 American Fisheries Promotion Act that amended the FCMA and changed its name to the MFCMA. The amended act included a "fish and chips" policy that raised fees and established stricter criteria for granting foreign fishing rights in U.S. waters. The needs of the fishing industry were also addressed through the 1980 Regulatory Flexibility Act, which required that fishery regulations be assessed for their potential to have significant economic impacts on small businesses. Under a 1983 Presidential Proclamation, the 200-mile Fishery Conservation Zone broadened in definition to become an Exclusive Economic Zone.[18] In 1987, the Commercial Fishing Industry Vessel Anti-Reflagging Act excluded from U.S. fisheries vessels built in foreign shipyards and limited foreign investment in fishing businesses—although loopholes in the law resulted in its having the opposite effect, leading some to call it the "Reflagging Act."

By the mid-1980s concern over the performance of the regional fishery management councils in conserving fishery resources led a National Oceanic and Atmospheric Administration "Blue Ribbon" panel to call for the separation of conservation and allocation decisions in fishery management. Under this recommendation, acceptable biological catch levels would be set outside of the regional council process by independent scientists to reduce the temptation to use local socio-economic factors as reasons to overfish.[19] This model of decision separation, while not applied directly within councils, is nevertheless the outcome in those councils that place high value on the recommendations of their scientific advisers. The North Pacific Council, for example, has exceeded the acceptable biological catch recommendations of its scientists only once.[20] This separation of conservation from allocation is used for Pacific halibut, where the International Pacific Halibut Commission sets the quotas and the councils allocate them among user groups.

By 1989, concern about overfishing led the National Marine Fisheries Service to revise the "602 overfishing guidelines," requiring overfishing definitions to be specified in fishery management plans for every fish stock. As fishery yields leveled off or declined nationwide, the social and economic stress on fisheries increased. It was clear to many that the commercial fishery growth rates of the late 1970s could not be sustained—that entry into fisheries would have to be limited. In addition, recreational interests in fishing were growing in number and influence. What was at the beginning of the decade a radical idea—that some people would be excluded from fishing—became by the end of the decade an accepted requirement in many fisheries. By the late 1980s the regional fishery management councils and the states had developed a wide variety of limited entry programs, but their effectiveness was constrained by being implemented piecemeal over many years, generally only after a fishery was experiencing serious problems and too many vessels had already entered the fishery. To implement these programs required combating the attitude described by one industry representative as "By God, if I invest in a boat and gear I should be able to go out and kill fish." Another now laments that "we missed the boat by not imposing across-the-board limited access schemes when we first implemented the FCMA." A number of fisheries remain open access today.

1990s: Contraction and Change

> Unfortunately, some still see fishing as the last bastion of industry where you live and die by your own abilities; where there is no role for government in your success or failure.
>
> Tom Hill, New England Fishery Management Council

By the 1990s domestic fishing fleets had developed to the point of full or overcapacity. To use the common expression, too many boats were

chasing too few fish. Even the better-functioning councils faced problems of expanding workloads and an increasing complexity of issues.
Just as fisheries were stressed, fishery management services were also
under pressure. Disputes between fisheries increased as they competed
for a share of the limited resource. Recreational fisheries continued to
increase in importance and visibility, demanding their share of fishery
allocations. Environmental interests, too, began to be a more visible
presence in management. In a landmark 1991 case, the Conservation
Law Foundation and the Massachusetts Audubon Society successfully
sued the National Marine Fisheries Service and the Department of Commerce over their failure to manage New England cod, haddock, and yellowtail flounder according to the mandates of the law.[21]

That same year foreign catching and processing were completely
eliminated from the Exclusive Economic Zone, achieving one of the
major objectives of the FCMA. Another objective of the FCMA was
realized as seafood imports continued to fall as a percentage of total
seafood supply.[22] By the mid-1990s, growing concern for the social and
cultural impacts of domestic conditions in the fisheries led to the
beginning of more formal treatment of social impacts.[23]

Public views of what is important in fishery management were also
changing. Our interviews identified several larger-scale societal and
environmental changes since the FCMA that influenced these views.
Coastal populations have increased and become more diverse in their
values and expectations for fisheries. Increasing numbers of people
desire recreational fishing opportunities. Larger populations in coastal
regions have also led to increases in environmental stresses that
degrade fishery habitat. Ocean regime shifts, such as El Niño, have low-
ered fishery productivity. The conservation community, inactive in fishery management in the early days of FCMA implementation, has
become an increasingly important presence in fisheries, calling public
attention to issues of bycatch, waste, and habitat.

The amended 1996 MSFCMA reflected these changing attitudes in
its concern for conservation and mandates to protect essential fish
habitat, rebuild overfished fisheries, reduce bycatch, and assess and
minimize management impacts on coastal communities. Steps toward
"Americanization," which began with restrictions on foreign fishing in
the American Fisheries Promotion Act and the Anti-Reflagging Act,
continued with the passage of the 1998 American Fisheries Act,[24]
which took further action to restrict the operation of foreign rebuilt
U.S. flag vessels—mostly factory trawlers—in the North Pacific.

By 1999 councils had been operating in the eight regions for 22
years, developing fishery management plans and plan amendments,

and allocating resources among user groups. By this time, participation in the council system had broadened to include a greater representation of the recreational fishing industry and environmental organizations on advisory panels and as council members.

Conclusion

> Open access is a deep-rooted tradition in fisheries. Because we part with it reluctantly and far too late in most cases, we're always chasing excess effort and trying to reduce overcapitalization.
>
> *Burr Heneman, marine policy consultant*

> The biggest problem with the Magnuson Act is that it didn't create a firewall between conservation and allocation questions.
>
> *Paul Brouha, formerly American Fisheries Society*

Fishery management is a product of its times and history. Management in the post-war period has emphasized fishery development and expansion, but must now make a transition to fishery stability, and in some cases, contraction. The ability of fishery management to make this transition is affected by the legacy of the past and also by the degree to which current management problems are successfully addressed. Experience has shown that management changes often lag behind changes in other parts of the fishery such as fishing technology, seafood markets, and public values.

Over these 22 years, the histories of individual councils have varied with their regional biological, economic, and social contexts. The eight councils have had varying degrees of success in blending the interests of fishery use and conservation into sustainable fishing practices. For some councils, decisionmaking has been outpaced by changes in fisheries. For others, fishery management has been more successful at maintaining effective management control. All councils have been criticized at one time or another for shortsightedness, conflicts of interest, and reactive decisionmaking. They have often been seen as captured by short-term interests at the expense of long-term sustainability.

Public attitudes toward fisheries have changed in the years since the FCMA was first implemented. The public no longer looks at marine ecosystems as simply food production systems. Much broader values for recreational use, aesthetic enjoyment, and biodiversity now apply. The 1996 amendments are in many ways a reflection of these changing public attitudes toward fisheries, and although some stakeholders resist the change, others are optimistic that the new requirements will help correct bad practices in both fishing and fishery management.

In Their Own Words*

*These statements reflect the personal views of interview participants and do not necessarily represent the positions of the institutions or organizations with which they are associated.

Anonymous Industry Sector Representative

An early failure in fisheries management, but one that undermines conservation even today, was Congress's decision in the early 1980s to limit the Commerce Department's authority to review actions by the regional fishery management councils. Granting increased autonomy to user group–dominated councils, while removing authority from the National Marine Fisheries Service's professional fishery managers, has compromised conservation and resulted in numerous fisheries failures, not the least of which was the New England groundfish fishery.

Anonymous Scientist Associated with the Fishery Management Council Process

We got here because of the historical paradigm of inexhaustibility, where people saw fisheries as a frontier and fishermen as buffalo hunters. There is a lot of glamour associated with people who matched their abilities against the difficult environment. The public doesn't see the fish the way they saw the buffalo. They see the boats in the harbor, fishermen, and coffee tables made of lobster pots. Somehow preserving the much romanticized industry of quaint fishing ports has overshadowed concern about the fish that are not visible to most people. It's been a very slow process by which reality has sunk in about the need to conserve the fish.

Anonymous State Fisheries Official

The council and the National Marine Fisheries Service never got together to work out what they wanted and how they would achieve it. The National Marine Fisheries Service thought it was in charge; the council wanted to be and thought it was. And the tension has been there ever since. Nobody wanted to invest in the process to sort out those relationships. We chose instead to invest in the short-term results and to concentrate on the details of substance. I don't think there has ever been a commitment by the National Marine Fisheries Service to make the councils work. At the same time, I don't think there's a recognition by the councils that the National Marine Fisheries Service has problems, too, that are imposed on them by external events and forces, and they deserve help in trying to resolve those problems.

Paul Brouha, Former Executive Director, American Fisheries Society

In 1976 we came to the realization that the system wasn't working and we, to some extent, correctly identified foreign fishing as the problem—get the exploiters out. And now the exploiters are our own people, and they're politically empowered, and we can't stop them! With the original Magnuson Act we created the new enemy, which happens to be politically enfranchised. The councils should be making allocation decisions, not conservation decisions.

Don DeMaria, Captain, F/V *Misteriosa*, Summerland Key, Florida

Expanded technology has been an important factor in fisheries management. With new technology, we've had rapid stock declines in the past 15 years. When you compare what we've got now with what we had—wooden boats, cotton nets—it was hard to do much damage. Change was so rapid, especially in the electronics field. The fish don't have a chance.

Graciela García-Moliner, Special Assistant to the Executive Director, Caribbean Fishery Management Council

In looking at what has happened, I think that cowardice has had a lot to do with it. Knowing that you have to implement measures but not taking the bull by the horns . . . letting things go for too long. Management of the nassau grouper is a good example; implementing management after the spawning aggregations are gone. But it's not just cowardice from the federal side, the state and users have responsibility as well. We definitely need better intra- and inter-agency cooperation.

Tom Hill, Chair, New England Fishery Management Council

The short-term costs to the industry were always, in my opinion, a determining factor in management decisions. The councils, in particular the New England Council, have tried to balance the social and economic concerns of the industry with mortality objectives. And, unfortunately, they've done a poor job of evaluating the cost-benefit risks relative to dislocation and the reality of continued overfishing. Short-term costs have always been viewed as greater than the risks of long-term failure. The council, with good intent at times, has half measured itself into failure. We've taken measures to stop the decline of a given resource but the measures have never been sufficient. All these small, incremental costs have caused economic hardship but have not fixed the problem.

Gerald Leape, Legislative Director, Ocean Issues, Greenpeace

One of the underlying problems that faces the federal agency charged with managing our nation's fisheries, the National Marine Fisheries Service, is that it has been given a conflicting mandate: to develop our fishing fleet while conserving our fish. That might have worked well in the past but now we're reaching a point where those objectives are in direct conflict. Unfortunately for the fish, the National Marine Fisheries Service, when faced with a conflict, has historically picked developmental goals over conservation goals. This has to change.

Alison Rieser, Director, Marine Law Institute, University of Maine

The basic problem in implementing the Magnuson Act early on was that the councils were given too broad a scope given that management by the industry itself was an untested idea. I believe in self-management with safeguards, but we jumped into it when there was virtually no history of industry involvement with the scientific side of fisheries management. Except for a few who had previously been [International Convention for Northwest Atlantic Fisheries] advisers, they didn't have time to become familiar with the science before they were required to make major decisions whether to believe it or not. Because stock assessments are estimates based on models and extrapolations, the results are easy for laypeople to reject. So the ease of exploiting scientific uncertainty, coupled with a lack of effective oversight of council decisions to disregard the warning signs, was a recipe for disaster.

Dick Schaefer, Chief, Intergovernmental and Recreational Fisheries, National Marine Fisheries Service

The passage of the FCMA was one of the most important events, if not the most important, in fisheries management. The creation of the councils was innovative, unprecedented, and far thinking. However, a decision made early on that councils not be staffed with federal employees has, in my opinion, not served the management process well. Indeed, it has contributed to the creation of two separate, and frequently adversarial, "camps," i.e., "us" vs. "them." Further, from the very outset, I have been concerned about council membership, i.e., a "mix" of state and federal fishery managers charged with stewardship responsibility for the nation's fisheries resources and persons from the private sector. The latter are frequently perceived as "foxes watching over the hen house." In my view, that perception has not helped to build council credibility and public trust in the "system."

Notes

1. Wenk, E. 1972. *The Politics of the Ocean*. University of Washington Press, Seattle, WA.

2. Wise, J. P., ed. 1974. *The United States Marine Fishery Resource*. National Marine Fisheries Service, National Oceanic and Atmospheric Administration. MARMAP Contribution, no. 1.

3. NMFS. 1977. *Fisheries Statistics of the United States, 1976*. NOAA, U.S. Department of Commerce.

4. Hennemuth, R. C., and S. Rockwell. 1987. History of Fisheries Conservation and Management. In R. H. Backus, ed., *Georges Bank*, 430–46. MIT Press, Cambridge.

5. Miles, E., S. Gibbs, D. Fluharty, C. Dawson, and D. Teeter. 1982. *The Management of Marine Regions: The North Pacific*. University of California Press, Berkeley.

6. Wenk, E. 1972. n. 1.

7. Magnuson, W. G. 1968. The Opportunity Is Waiting . . . Make the Most of It. In *The Future of the Fishing Industry of the United States*, 7–8. University of Washington Publications in Fisheries, n.s. vol. 4.

8. Commission on Marine Science, Engineering and Resources. 1969. *Our Nation and the Sea: A Plan for National Action*. U.S. Government Printing Office, Washington, DC.

9. Wise, J. P. 1991. *Federal Conservation and Management of Marine Fisheries in the United States*. Center for Marine Conservation, Washington, DC.

10. Magnuson, W. 1997. The Fishery Conservation and Management Act of 1976: First Step Toward Improved Management of Marine Fisheries. *Washington Law Review* 52(3) (July): 42–50.

11. Public Law 94-265, 90 Stat. 331 (codified at 16 U.S. C.A. sections. 1801–1882 (West Supp. 1977)). The Fishery Conservation Zone was later renamed the Exclusive Economic Zone.

12. Wise, J. P. 1991, n. 9.

13. Alperin, I. M., C. H. Lyles, and J. P. Harville. 1977. *A Report to the Congress: Eastland Fisheries Survey.* Atlantic States Marine Fisheries Commission, Gulf States Marine Fisheries Commission, and Pacific States Marine Fisheries Commission.

14. Buck, E. H. 1995. *Social Aspects of Federal Fishery Management.* Congressional Research Service, Washington, DC.

15. NMFS. 1996. *Our Living Oceans. The Economic Status of U.S. Fisheries.* NOAA Technical Memorandum NMFS-F/SPO-22, December.

16. Ibid. Created in 1970, the fund deferred payment of federal income taxes if the money was used to construct or refurbish a vessel.

17. Ibid. Created in 1973, the program provided loans to finance up to 87.5 percent of the cost of vessel or shoreside facility construction or reconditioning.

18. Presidential Proclamation 5030. President Ronald Reagan.

19. Anon. 1986. *NOAA Fishery Management Study.* NOAA, Department of Commerce, Washington, DC.

20. Clarence Pautzke, executive director, North Pacific Fishery Management Council. Personal communication.

21. *Conservation Law Foundation v. Mosbacher*, 966 D. 2d 39 (1st Cir. 1992).

22. NMFS. 1992. *Fisheries Statistics of the United States, 1991.* Current Fishery Statistics No. 9100. NOAA, Department of Commerce, Washington DC.

23. NMFS. 1994. *Guidelines and Principles for Social Impact Assessment.* NOAA Technical Memorandum. NMFS-F/SPO-16, May.

24. 105 Pub. L. 277 Omnibus Consolidated and Emergency Supplemental Appropriations Bill for Fiscal Year 1999.

PART III

The Present

*W*e're going through a period of rapid transition in fisheries management; from developing to protecting and sustaining fishery resources and ecosystems.

Dayton L. Alverson,
Natural Resources Consultants, Inc.

CHAPTER 3

Assessing Fishery Productivity

> We need to ask, what we want out of these fisheries? A viable fish-
> ing industry or a blue zoo?
>
> *Rod Moore, West Coast Seafood Processors Association*

Productivity. It seems such a simple word, but is it? The dictionary definition of "productivity" is "yielding useful or favorable results."[1] The
fishery definition is more complicated. Favorable results must have biological, economic, and social dimensions because fisheries have those
dimensions. Fishery productivity depends not only on the ability of fish
populations to produce fish, but also on their ability to provide economic value and social benefits like recreational opportunities and the
satisfaction of knowing fish exist.

The MSFCMA requires that fisheries be managed to provide "the
greatest overall benefit to the Nation, particularly with respect to food
production and recreational opportunities,"[2] subject to stocks being kept
at levels that can produce maximum sustainable yield, the maximum
average yield over the long term. This does not mean that fisheries must
always produce the highest yields possible, but only that stocks must be
kept at levels such that they can produce highest yields. The yield that
produces the greatest benefit, called "optimum yield," may not be the
largest weight of fish. Optimum yield could occur in smaller catches if
fishing costs are high, if the market absorbs only small quantities, or if
people value smaller sizes of fish. Before the 1996 MSFCMA amendments, the law defined optimum yield as the maximum sustainable yield
as modified by social, economic, or ecological factors. Under this definition, "as modified by" could mean greater than or less than maximum
sustainable yield. The 1996 amendments replaced "as modified by" with

39

"as reduced by" to guarantee that optimum yield could not be set at levels above maximum sustainable yield.[3]

The appropriateness of using maximum sustainable yield as a management target has been widely debated,[4,5] but it is clear that federal law intends there to be an upper limit on fish catch based upon the capacity of stocks to renew themselves. The use of maximum sustainable yield as that limit establishes biological productivity as the primary contributor to fishery benefits. But it is not the only contributor. What makes a fishery productive is that it creates economic and social, as well as biological, benefits, and complications arise because these three types of benefits are not always maximized at the same stock levels. Some stakeholders remind us that benefits to the nation are to be found in more than just food production and recreational fishing, "not just putting fish on the table or on the hook." Benefits also derive from enjoying but not killing fish, as from recreational diving on coral reefs, or from "just knowing that the fish are there," a value often expressed in support for fishery programs of conservation organizations.

What Is Productivity?

Biological, economic, and social factors all influence how we define and measure productivity. The measure of productivity is never static, but must continually adapt to changing conditions.[6]

Biological Productivity

> It's not a complicated issue. Climate change is a complicated issue
> . . . leaving enough juveniles in the water to grow up to lay eggs is
> not a complicated issue.
>
> *Carl Safina, National Audubon Society*

> In a mathematical equation, you can set the variables equal to zero
> to work out the answer. But you can't do that with [management]
> decisions because one of the variables is unknown. . . . Environ-
> mental variability is the unknown factor.
>
> *Graciela Garcia-Moliner, Caribbean Fishery Management Council*

The National Marine Fisheries Service publication *Our Living Oceans*[7] represents the biological productivity of fishery resources in three concepts: (1) recent annual yield, the average catch over a three-year time period; (2) current potential yield, the potential catch based on current stock abundance and ecosystem considerations; and (3) long-term potential yield, the maximum long-term average catch that can be achieved—a concept analogous to maximum sustainable yield.

What a fish stock produces depends on how fast it grows, reproduces, and dies. If a stock is to remain productive over time, the rate at which fish are killed must not exceed the rate at which they renew themselves through reproduction and growth. Fish die from natural, human, and environmental causes. Fish eat each other and are killed by environmental conditions that affect the quality and quantity of their habitat. Fish are killed in targeted fishing and as they are caught incidentally— as bycatch—to targeted species, dying by direct injury from the gear, from injury during handling and release, or from the "ghost fishing" of lost gear. Fish are also killed indirectly from gear disturbances to habitat or from fishing practices that alter the mix of fish stocks. Ocean conditions affect the availability of food for fish and so also affect their rates of growth and survival. The primary concern of fishery management is the fish that are killed by fishing, although management now has the additional responsibility of protecting fish habitat from the harmful effects of fishing.

If actual catches are compared to their potential, it is clear that fisheries are not as biologically productive as they could be. The National Marine Fisheries Service reports that in 1995 the recent annual yield of U.S. marine fisheries was 7.2 million metric tons, or about 63 percent of its long-term potential.[8] The primary reason for the shortfall is overfishing—catching levels of fish that reduce the populations to unacceptably low levels. Too much is being taken too fast or too early from fish populations.

Fishery landings statistics reveal only a part of fishing mortality— that portion of the catch which is brought onshore. Fish are also caught and released in recreational fisheries or are caught and discarded in commercial fisheries because they are unmarketable, too small, exceed a catch limit, or are prohibited by other regulations. Some of these released or discarded fish also die. From the perspective of biological productivity, it is the total fishing mortality—landed catch plus bycatch mortality—that matters. In the past, when fishing pressure was low relative to the size of the fishery resource, bycatch was unimportant to biological productivity. Fish stocks could tolerate the small mortality that resulted. But fishing pressure is no longer low, and the quantity of fish killed as bycatch is a significant factor for many stocks. Although no estimate is available for the United States, worldwide it is estimated that of approximately 84 million metric tons of marine fish landed each year in commercial fisheries,[9] another 27 million metric tons are caught incidentally to targeted species and discarded dead.[10] An additional, but largely unquantified, amount of fish die as both catch and bycatch in recreational fisheries. The evidence suggests that these levels of fishing mortality are unsustainable and that we are above the maximum sustainable level of many fisheries' biological productivity.[11] Some fisheries, such as the Alaska groundfish fishery, have observers onboard vessels to measure and monitor bycatch and discards and count bycatch as part of the total allowable catch.[12] But in most fisheries, the quantity of bycatch remains an estimate.

Economic Productivity

> The bottom line is that there is an enormous amount of potential societal benefit from most U.S. fisheries that is being depleted by depleting the stocks.
>
> Doug Hopkins, Environmental Defense Fund

> An economically viable fishery is one with a future, with young people entering.
>
> Richard Young, F/V City of Eureka, Crescent City, California

The economic productivity of fisheries builds on biological productivity, but is not determined by abundant numbers of fish alone. Economic productivity is also determined by how the fish are caught and used and the extent to which their use contributes to human betterment. One way to look at economic productivity is to consider how efficient a fishery is—how the revenues generated by the fishery compare to the costs of earning those revenues. Efficiency is highest where the difference between revenues and costs is at its greatest—where net revenues are at their maximum level. Efficiency is also high when fishing is cost-effective—when a specified amount of fish is caught at the least possible cost.

Economic productivity also relates to the extent to which fisheries contribute economic benefits to individuals or communities through employment, processing, sales of equipment and supplies, transportation, and other related businesses such as restaurants and hotels used by anglers. This type of productivity includes not only the size of the benefits, but also how they are distributed. In rural areas, for example, fisheries may be the major—or only—source of employment. Economic productivity in these fishing-dependent regions takes on heightened importance in its wider economic impacts.

Economic productivity is also measured by the return on investments made by fishing businesses. But there is a difference between the efficiency of individual fishing businesses and the efficiency of the fishery as a whole. For an individual business, efficiency is a matter of combining all the pieces of fishing effort—boat, gear, electronics, fuel, ice, skill, labor—in a way that produces the greatest profit, or private net benefit. For the fishery and its public owners, efficiency is a broader matter—it is where net public benefits are the greatest. Public benefits include the employment, material well-being, tax revenues, and general economic activity created by fisheries. But just as fisheries generate benefits to the public, they also forgo value and create costs. Value is forgone when fish are wasted as bycatch, when scarcity makes it cost more to find and catch fish, when recreational opportunities shrink, and when management costs rise. All the fishery management activities of data collection, analysis, decisionmaking, and enforcement—paid for by the public—increase as fisheries are more intensely used. The magnitude of public benefits and costs changes over time with changes in markets, public preferences, management approaches, and ecological conditions, and so the economic productivity of fisheries also varies.

Social Productivity

> Social benefits include . . . watching a salmon run or feeding fish from a bridge, as my four-year-old son likes to do.
>
> *Peter Fricke, National Marine Fisheries Service*

> If there were simple fixes, things would already be fixed; no one likes the present environment. But things are socially complex.
>
> *Tom Hill, New England Fishery Management Council*

The social productivity of fisheries relates to objectives such as achieving fairness in income distribution and diversity in scales of fishing operations, providing opportunities for recreational and subsistence fishing, sustaining coastal communities, maintaining fishing culture, and transmitting knowledge. Because ideas about fairness, scale, culture, and community vary by region, so also do social objectives. The 1976 FCMA recognized these regional differences in fisheries when it set up fishery management as a decentralized regional system.

Social productivity depends on the biological and economic productivity of fisheries. Employment, income distributions, fishing opportunities, and the resilience of coastal communities all stem from the flow of economic activity associated with the fishery's biological production. The factors we typically consider to be social dimensions of fisheries include those things that contribute to quality of life—physical and mental health; absence of social distress from violence, crime, or substance abuse; access to desirable employment; aesthetic appreciation; and recreation. The biological and economic productivity level of fisheries affects the social well-being of individuals, families, fishing communities, and fishing regions and the quality of fishing as a way of life.

Because social objectives for fisheries are not always the same as economic objectives, ideas about social and economic productivity can be in conflict, requiring managers to make trade-offs between them. For example, a fishery with a few large producers might be economically efficient but might be considered by some as unfair because it employs few people.

At present there is less agreement about how social productivity should be considered or measured in fishery management than about either biological or economic productivity. It is often taken for granted as a background issue until it is significantly diminished. The diminishment of social productivity is more likely to be observed and measured when fisheries are in poor condition and showing signs of social distress. In part this results from the difficulty of separating the social issues of fisheries from the social issues of the general society.

Managing for Productivity

> I believe that social and economic benefits will flow from biologically
> and ecologically healthy fisheries. What's good for the oceans over
> the long term will tend to be good for humans over the long term.
>
> Lisa Speer, Natural Resources Defense Council

What determines fishery productivity depends on the objectives for the fishery. Most fisheries are not pursued for a single commercial, recreational, or subsistence purpose, and this means, as an academic stakeholder observes, that there is no single best benefit to be derived from a fishery—there are only different packages of benefits. The choice of the package is up to fishery managers, and each has its own implications.

Atlantic billfish (marlin, sailfish, and spearfish) are a good example. They are an international "high seas" fishery, with the U.S. portion of the quota managed by the National Marine Fisheries Service for the objective of maximum sustainable recreation, rather than maximum sustainable yield.[13] Atlantic billfish are reserved exclusively for recreational fishing and are not allowed to be sold. Most recreational billfish fishing is catch and release, so the objective of maximizing recreational opportunities is best accomplished by keeping fishing mortality as low as possible so that fish can be caught and released as many times as possible. It would be simple if this were the whole story—the biological productivity of billfish in recreational use would be high. But some of the fish that are caught in the recreational fishery do not survive release. Large numbers of billfish are also caught incidentally and unavoidably on commercial longline gear set for swordfish that have different management objectives. In addition, the fisheries of other countries cause billfish mortality. The result of all these factors is that billfish are overfished despite being managed for nonconsumptive use in the United States, illustrating how complicated a package of benefits can be.

Because the number of possible benefits from a fishery is large and the failure to prioritize those benefits is common, it is difficult to determine how many opportunities are being lost or how much benefit is being forgone by the way fish are managed. This is clearly the case in the Gulf of Mexico red snapper fishery. In this fishery, the large adult fish are sought by commercial fishermen using a variety of gear types and by recreational fishermen using hook and line. Juvenile fish are caught by these fishermen but are usually released because of regulations—some of these released fish die. In addition, large numbers of juvenile red snapper are caught incidentally in commercial shrimp trawls—nearly all of these fish die. The result of the combined bycatch mortalities is that the stock is so severely overfished that even a complete prohibition of the large-fish commercial fishery would not likely be enough to rebuild the stock.[14] Reducing the shrimp trawl bycatch would also be necessary.

As with most fisheries, there are many perspectives on appropriate objectives for red snapper. The requirement for bycatch reduction devices to allow escapement of red snapper has imposed additional costs on the shrimp fishery. Shrimp trawlers argue that the cost to the shrimp fishery to save the commercial red snapper fishery is too high. The red snapper fishermen argue that shrimpers should not be allowed to waste these fish to the detriment of the red snapper fishery. Managers contend that there is a resource conservation issue apart from the allocation of red snapper among fishermen. At present these different perspectives are not assessed on the basis of benefits that could result from managing for one objective over another, but if the potential productivity of the red snapper fishery is to be realized, relative benefits and the trade-offs between them must be analyzed.

Social and economic objectives have the potential to conflict with the clear mandate to maintain stocks at levels that are capable of producing maximum sustainable yield. Fish managed for the benefit of a commercial fishery can often be of different size or volume than fish managed for the benefit of a recreational fishery. Even while appearing to be biologically healthy, fisheries may have significant unmet social or economic potential. Biological success is not necessarily economic or social success, and that is why relying on a single measure of productivity is inadequate to understanding the productivity of the fishery as a whole. One person we interviewed points to the North Pacific halibut fishery as an example. Prior to the adoption of individual fishing quotas, management was biologically successful, but the fishery itself was economically and socially unsuccessful under an intensive race for fish. Fishing periods lasted only 24 hours; halibut were landed in huge gluts and stored frozen. Fresh fish were only intermittently available, prices were low, and fishing safety was compromised. Since the adoption of management by individual fishing quotas, improvements in all these areas have been achieved as social, economic, and biological objectives have been better integrated.[15]

While many in the North Pacific halibut fishery are delighted with this outcome of the individual quota program, it is important to point out that making the change to individual quotas did not make everyone happy. Social and economic objectives still conflict for some who feel they received too small a quota share and for others who feel that having to buy their way into the fishery is unfair. Still others object to the restrictions built into the program.[16] These widely different views of the program illustrate the difficulty managers face in balancing ideas about biological, economic, and social productivity.

The Alaska groundfish fisheries are another example of fisheries that are generally considered to be biologically healthy and yet have unmet economic potential. Total catch has been relatively stable, and the domestic fishery has flourished.[17] But the full potential value of Alaska groundfish has not been met because of excess fishing capacity in both factory trawlers and catcher boats. This fishery is not overfished. It is conservatively managed by the North Pacific Fishery Management Council. But the socioeconomic pressures of overcapacity mean that it costs more to catch and allocate the fish than necessary. Biologically healthy resources and stable high catches are not sufficient to ensure the greatest economic productivity.[18]

Management also affects productivity when it allocates fish among user groups. All techniques of fishery management create winners and losers, even those that employ biologically based conservation measures. Closures of spawning grounds mean fishermen must find other times and areas to fish. Closures of nearshore areas for marine protected areas may restrict the access of recreational anglers to fisheries. Minimum fish-size regulations take fish away from fishermen who fish in areas with concentrations of small fish. A trawl mesh size that allows the escapement of smaller species incurs a cost to fishermen who fish species of varying sizes. Open access total, allowable catches favor fishermen who can catch the fish before their competitors. Measures to spread total allowable catches out over time take fish away from those who have invested in the capability to catch fish quickly. While generally thought of as "passive" or "indirect" management measures, these traditional regulations also affect economic productivity.

Market-based management measures, such as individual transferable quotas, are directed at increasing economic productivity and bringing it into line with biological productivity. The ability to sell or lease an option to catch a share of the total allowable catch allows fish to be caught by those who place the highest value on that share. Individual owners can choose whether to hold quota share through a period of reduced fishing, buy additional share, or sell their share and invest in another endeavor. They can fit quota share to their targeted catch mixes. Without marketable shares, as one stakeholder points out, fishermen have no financial equity in the fishery and cannot sell out and leave the fishery during times of low productivity. In these cases regulations will usually spread out available fish among all participants, creating a situation where "we all go broke together." But even with marketable shares, people are vulnerable to low biological productivity because the value of a share is determined by how much fish it is likely to represent.

Maintaining Productivity Over Time

I was raised and taught to take what we needed from nature but to leave an adequate amount for the future to sustain oneself and nature as well.

Jim Harp, Quinault Indian Nation

Sustainable fisheries means managing for the future as well as the present. This should not be a difficult standard to achieve if the political will is there.

Jim Gilmore, At-sea Processors Association

When people talk about productivity, they inevitably talk about sustainability, and the idea of sustainability links people to fish. Because fishery systems include both fish and the people who depend on them, and because there is uncertainty about the future, the idea of sustainability must include the option to maintain the productivity of fish populations and their habitat, as well as humans and the communities in which they live.

The stakeholders we interviewed use the term "sustainable fisheries" to mean a number of things. To some, it is little more than a catch-all term to mean anything—sustainability is "in the eye of the beholder," one says. But stakeholders from all sectors see sustainability as meaning leaving enough fish for the future. To most, it also means maintaining a relatively high level of productivity "from generation to generation" without waste or harm to the ecosystem. The question is, High for whom? And in variable systems, can production always be high?

The different ideas about what is "enough" and "who gets what" become the source of contention that creates political difficulty for managers. Many people we interviewed emphasize that a sustainable fishery would have to include economically healthy fishing businesses as well as biologically healthy fish stocks. Several point out that by maintaining fish populations at healthy levels, the businesses and communities that depend on those fish are also maintained. Without long-run sustainability, other benefits are not possible, and arguments about how to allocate fish become meaningless.

Balance is the key to sustainability for many—balance between today and tomorrow, use and conservation, biology and economics, communities and individuals, variability and stability. It is complicated because the appropriate balance is different for different people. Fishery managers are required to put conservation first, and then decide how to divide the fish among all the competing uses. The difficulty for managers is that they must address conservation and allocation in that order, but they are often prevailed upon to let allocation needs determine their conservation decisions. As a scientist notes, the weakness of management is that it fails to maintain the priority of conservation because it "takes into account only the short-term interests of those who are currently active in the fishery, instead of the general public, now and in the future."

Achieving sustainable fisheries, many note, requires accommodating the natural variations of the ocean. But managing for constancy in a variable ocean environment is fruitless. Part of the task of fishery management is to balance between years of high and low abundance in the fishery—to manage for a level of use that allows the fishery and fishing communities to absorb the low abundance as well as the high. "To some, sustainability means sustaining everything in the position they're in now—I don't buy into this," says one long-time participant in fishery management. "Efforts to protect everyone and their stake in the fishery make sustaining fisheries at a profitable level impossible."

Stakeholders also point out that a fishery can be sustained at many levels—even at very low levels of productivity—as long as what is removed does not exceed what the stock can produce. Even so, many observe, it is hard to think that sustaining fisheries at levels below what will produce maximum sustainable yield is optimal in any sense. A stakeholder argues that a benefit of maintaining fish stocks at levels capable of producing maximum sustainable yield is that this is where they provide the most options. At this level, he asserts, fishing and its economies can be supported, buffers can be provided against risk and uncertainty, biodiversity and ecosystem health can be maintained, and the choices about types of use can be maximized. But there will still be trade-offs and choices to be made among competing uses of fisheries. Reducing fish stocks below levels that could produce maximum yields means losing some of the options for choice.

Looking beyond fish stocks to the sustainability of ecosystems is an idea that is often acknowledged, but rarely implemented. One researcher notes that because of ecosystem variability, sustainability needs to be measured by the health of the entire ecosystem rather than by the health of individual species—by healthy patterns of ecosystem variability. Others point out that although single-species management can be effective for maintaining populations of individual species—for example, striped bass in the Chesapeake Bay—other organisms may be affected through changing interactions of fish at different trophic levels, like striped bass and crabs. Fishery management plans are now beginning to take some ecosystem factors into account, such as the effect of fishery removals on forage for marine birds and the effect of fishing gear on habitat.

Some ideas for ecosystem management involve setting aside areas protected from human use. The 1998 report to Congress of the Ecosystem Principles Advisory Panel[19] recommends that fishery managers consider and evaluate the potential benefits of marine protected areas for promoting ecosystem-based management. Marine protected areas can range in size and degree of exclusion. For example, some may prohibit all forms of fishing and nonfishing use; others may restrict just commercial and recreational fishing gear; others, commercial gear only; and still others, specific types of commercial gear. Prohibitions in some protected areas may remain in effect year round, while other marine protected areas may restrict activity only during certain times; for example, when fish are spawning.

The management question is how to balance the benefits of area-based protection against their costs. The challenge facing fishery managers is to ensure that fishing—commercial, recreational, and subsistence—allows ecosystem health to be maintained. This challenge is summarized in one of our interviews as maintaining the abundance and complexity of the natural variability of ecosystems, while also continuing to provide fishermen access to fisheries resources. We are not saving fish for the purpose of saving the fish, but for human use instead, according to one fishery manager.

In acknowledging the difficulty of managing at the ecosystem scale, several stakeholders also think that the answer to maintaining healthy ecosystems might be found in what we already know and practice. We may not need to develop complicated and information-intensive approaches to ecosystem management. The easier alternative may be a more conservative application of the rules for managing individual stocks. Better management would keep stock levels higher, making them more available as food resources to other fish and marine mammals and keeping the entire ecosystem healthier.

The experience of the past 20 years makes it clear that to manage fisheries sustainably in the long run requires being more cautious in the short run. As a result, more management decisions incorporate a precautionary approach. This approach recognizes that changes in fisheries systems are only slowly reversible, difficult to control, not well understood, and subject to changing environment and human values.[20] It says that where the likely impact of fishing is uncertain, managers should be conservative and not "kill the goose that laid the golden egg," as described by one stakeholder. The MSFCMA, although not explicitly using the term "precautionary," nevertheless reflects it through the modified definition of optimum yield, the definition of overfishing, and the requirement to rebuild overfished stocks.

In the past, the health of fish stocks has often been sacrificed to short-term economic and social concerns, leading eventually to more severe economic and social losses. Finding the appropriate balance between biological, economic, and social benefits is a question that is on many minds. What short-term economic and social sacrifices are appropriate to rebuild fish stocks to maintain long-term benefits? This is one of the most troublesome questions of fishery management.

Conservation is a process of saving and investment.[21] People vary in their willingness and ability to save and invest—those whose cash flow will not even pay off existing debts are not likely to put aside savings or invest for future returns. Investments in fishery conservation are no different in this regard. People are more likely to want to save fish for the future—invest in future fishery productivity—when the fishery is economically healthy, which explains, at least in part, the objections raised against forced investments in fishery conservation when fisheries are under stress. The magnitude of the investment in conservation or rebuilding and the sacrifice it requires depend on the status of the stocks as well as on the economic condition of the fishing business. The healthier the stocks and the more profitable the vessels, the more modest is the investment required. With many fisheries, conservation policies have failed to be implemented early enough to require only modest sacrifice. Instead, management has often waited until the cost of taking action has become economically intolerable and politically unacceptable.

Many of the people we interviewed acknowledge the difficulty of asking fishermen to sacrifice present income for uncertain future gains. Even those stakeholders who emphasize the need for conservation are generally sympathetic to the human pain and economic hardship associated with rebuilding depleted stocks. Rebuilding stocks often requires a time period longer than most people will be involved in the fishery. The longer the rebuilding period, the clearer the trade-offs between individual pain and societal gain become—between one generation's costs and another generation's benefits. The problem is that those who pay the costs of cutbacks do not expect to be the ones who reap the benefits.

Sometimes people in the fishery are required to operate under more restrictive regulations in response to stock declines not caused by fishing. Many stocks change abundance as a result of natural events like El Niño or population cycles. A common response to this situation is to protest that "we didn't cause this downturn, so why should we cut back?" But just as droughts caused by changing weather patterns require restrictions on water use, so do natural fluctuations in fish stocks require increased restrictions on fish catches.

There are many examples of fisheries that have suffered from short-sighted management. An environmentalist cites the example of spiny dogfish in New England and the Mid-Atlantic. "We encouraged the development of the spiny dogfish fishery without foresight and without a fishery management plan," she laments. "We should have learned our lesson from the intense fishing of sharks . . . dogfish will take 10 years to recover—the fishery will probably have to be shut down. If we had managed this fishery from the beginning, we could have avoided putting fishermen out of business." As members of all sectors recognize, the loss of future opportunities as a result of the failure to anticipate and prevent unwanted outcomes plagues fisheries nationwide.

But there are also examples of fisheries where, once the problems of short-sightedness have been observed, corrective action has been taken. North Pacific halibut, arguably a far more productive, sustainable, and safer fishery under individual fishing quota management, is one example.[22]

Lost Opportunities

> We must give society the opportunity to choose. As it is now, we
> are constantly reducing our options.
>
> *Ken Hinman, National Coalition for Marine Conservation*

American fisheries represent many lost opportunities. In some, like
Pacific salmon, the losses are obvious, taking the form of endangered
species listings, overfishing, displaced commercial fishermen, reduced
recreational opportunities, diminished tribal fisheries, habitat degradation, and social conflict.[23] Biological depletion represents an obvious
deficit in contributing benefits to the nation. In other fisheries, losses
are much less obvious. As discussed earlier, even biologically productive
fisheries may not be achieving their full social or economic potential
because of the way we have chosen to manage them. Just as productivity has biological, economic, and social dimensions, so too does lost
opportunity.

Lost biological opportunities abound. Thirty percent of the 300 U.S.
marine fish stocks for which the status is known are overfished.[24] Overfishing occurs in two forms: The first is "growth overfishing," when fish
are caught before they grow to the desired age or size. The reproductive
capability of a stock may not be harmed, but the animals are caught at
a size far below their maximum potential size. The second is "recruitment overfishing," when fishing reduces the spawning stock size to a
level at which young fish are not added to the catchable population at
the desired rate. This can result from fishing targeted at the stock or
from fishing with wasteful, nonselective practices. Either way, the
reproductive capability of the stock is harmed.

Maximum sustainable yield is the benchmark for overfishing definitions now in place. This means that a stock may also become overfished,
even at very low levels of fishing if environmental factors cause stocks to
fall below levels that can produce maximum sustainable yields.

Lost economic opportunities are also widespread and are closely
linked to lost biological opportunities. Both growth and recruitment
overfishing result in yields that are smaller and often less valuable than
they could be. In recruitment overfishing, yields are smaller because the
size of the spawning stock is reduced below the level needed to produce
maximum sustainable yield. With growth overfishing, yields are smaller,
but the economic effect is a little more complicated. Some fisheries, such
as little-neck clams, herring for roe, and sardines, are economically more
productive when the animals are caught smaller because consumers prefer them that way. But in most fisheries, growth overfishing results in far

fewer benefits than could be produced, even when the fishery appears to be healthy.

The American lobster fishery, for example, has had record landings in recent years and shows no obvious signs of recruitment failure, although the scientific advice is that the lobster resource is overfished and at high risk.[25] Lobsters are taken at an average weight of less than 1.5 pounds, when they have the potential to grow to 30 or 40 pounds and gain approximately 40 percent in weight per year. Overfishing means that potential benefits from the fishery are money just "left on the table," and as both fishermen and consumers become accustomed to a smaller size selection over time, the losses become less visible. The size mix of lobsters that would produce the greatest economic benefit is a question that could be analyzed, and management could be targeted at this mix. One stakeholder remarks about fisheries in general that "just in economic terms, we could probably generate more than twice the present value of our fisheries, without having to make enormous trade-offs in achieving those benefits."

Recreational fisheries also represent lost opportunities because their total economic value is often unestimated and not fully recognized by management. This is in part because some of the values of recreational fishing relate to enjoyment and are not market related. But it is also in part due to the relatively new systems of accounting for the amount and types of recreational use and the economic benefits generated by anglers as they purchase equipment, supplies, boat trips, and other goods and services in coastal communities.

Lost opportunities can also result from overfishing that is economic in nature. Economic overfishing occurs when the fishery has developed beyond the point of earning the maximum net benefits. The immediate cause of economic overfishing may be either growth or recruitment overfishing, but the underlying cause is overcapacity—an amount of vessels and gear that exceeds the amount needed to catch fish at the least cost. A 1997 study of U.S. commercial marine capture fisheries estimated that they lose $2.9 billion per year in potential revenue as a result of overcapacity.[26]

Many of the reasons why overcapacity exists have their roots in the traditional approach to managing fisheries. When fishermen have no assurance that fish they leave uncaught today will be available to them tomorrow, they participate in a "race for fish," trying to catch as much as they can before their competitors do. This is what has become known as the "tragedy of the commons." Under these conditions fishermen use more capital and technology than would otherwise be necessary because even though the fishery is overcapitalized, they see their business as undercapitalized, requiring further investments to keep themselves competitive. Bigger, faster boats, more sophisticated electronics, and better gear may temporarily increase individual catch. But, over the long term, the total catch is the same if limited by a quota and is achieved at greater cost and sold at lower value. Fishing incomes are lower, and fisheries are less stable.

Overcapacity hurts the productivity of management as well. When capital and technology are increasing rapidly, the effectiveness of fishing effort increases, making it difficult for fishery regulations to keep pace. Additionally, debts associated with heavy investment in fishing capital create pressure on the fishery management system to set total allowable catch at risky levels. One environmentalist notes: "If there is an enormous overcapacity problem driving fishermen to overexploit, we'll be unlikely to achieve biological sustainability." A research scientist adds: "Resource agencies have been too preoccupied with preserving the fish stocks and have generally ignored the problem of economic overfishing." Overall, a shortage of economists on professional staff limits the ability of agencies to conduct these analyses. One stakeholder compared the strain on management resources to the race for fish. The race for free management resources—when those requesting management services don't have to pay—results in the same kind of overuse.

Fisheries represent many lost social opportunities. Some are related to a particular fishery, but others are general to all fisheries. Many are expressed in terms of the impact of fishery regulations on fishing communities. Reduced quotas and other measures required by declining fish stocks can result in fewer jobs on boats, in processing, and in marketing. They can also mean lower revenues for ancillary businesses such as ice suppliers, welding companies, and cafes. Job losses or restrictions in fishery participation can lead to sharpened divisions among community groups, thereby diminishing the sense of community cohesion and the capacity for facing new challenges. Increased competition for limited resources and profits may change patterns of work and pose new strains for family and community life, as when fishermen must stay at sea longer or travel to more distant fishing grounds. Work satisfaction suffers, as well as the opportunity to transmit the fishing culture to new generations. Some members of fishing communities cite the "terrible impact" of inefficient management and overcapacity. Others complain about the lack of community orientation in fishery management.

Perhaps as a reflection of the failure to prevent social losses or to specify social objectives for management, many social issues have been addressed by Congress in response to special interests. Since 1996, for example, laws have mandated a reduction in the number of factory trawlers in Alaska, prohibited factory trawler participation in the New England herring and mackerel fisheries, and placed a four-year moratorium on starting new individual fishing quota programs. These examples imply a mixture of social objectives that includes an emphasis on smaller-scale production, maintaining access to fisheries, and the distribution of fishery benefits.

Several stakeholders note the lost opportunity for fishery management to be more effective. Losses exist in many forms, including the failure to take the lead in developing innovative management approaches and—through the moratorium on the development of new individual fishing quota programs—to allow the market to help make the transition from overcapitalized to more efficient fisheries. Council management is blamed by some stakeholders for being slow to capture opportunities to educate the public about fisheries and to develop a broader ecosystem understanding. This raises the question of the appropriate educational role for councils. Many councils freely and widely distribute information about fisheries and fishery management actions. Is there a broader and more aggressive educational role that people have in mind?

Councils are also criticized by some for failing to control adversarial politics and for missing opportunities to promote cooperation among user groups. But others point out that councils are intended to be the arena in which the politics of different interests is worked out. Perhaps the point of those commenting is more that the toll exacted by crisis management, when time is spent on a sequence of crises, limits opportunities to address long-term objectives.

An assessment of the lost opportunities in commercial fishing is reflected in lower levels of optimism about a fishing future among the stakeholders we interviewed, as well as in a sample of commercial fishermen participating in a recent *National Fisherman* poll.[27] Of 3,500 fishermen polled, 81 percent said they were either "fearful" or "very fearful" in their outlook on the future of commercial fishing. Seventy-nine percent said they would not recommend fishing as a future occupation for their son or daughter. One fishing industry member we interviewed remarks that "folks are not talking up fishing to their children or their friends the way they used to." In consequence, some note, many young people have lost any thought of growing up to be part of a fishing community. Several industry members also explain that as fishing profitability declines, it gets harder to find high quality crew labor because crew wages are a share of vessel earnings.

If, as many stakeholders say, fisheries are rife with lost opportunities, the challenge facing the fishery management system, and society as a whole, is to regain opportunities to make fisheries productive. The situation will not be corrected by simply finding some yet-to-be-discovered vast fishery resources. "There are no new, or at best very few, opportunities left within the traditional fisheries," says one fishing industry representative. Opportunities will not be realized from the development of additional fisheries, but from improving the productivity of existing ones.

Conclusion

Healthy stocks will maintain healthy communities, healthy economies, and healthy fisheries.

Dick Schaefer, National Marine Fisheries Service

There is a fundamental problem with such objectives as maximum benefit: they push us toward, or over, the line of sustainability and incline us to override our uncertainty.

Burr Heneman, marine policy consultant

Fisheries are not as productive as they can be—biologically, economically, or socially. Although stakeholder views vary regarding which aspects of productivity are most important to successful fisheries, all agree that the management system at present is not structured to maximize productivity in any sense and that this must be corrected before additional opportunities are lost. Most fishery stakeholders recognize the importance of finding a balance between social, economic, and biological productivity. These three components will be much easier to balance over the long term and when the trade-offs between each are explicitly addressed.

Stakeholders of all different types are concerned about strengthening conservation in fishery management, although they have different ideas about how best to do this. Many indicate the need to develop practical definitions and measures of sustainable fishing that include buffers against overfishing, adjustments for uncertainty, and indicators of ecosystem health. It is clear that many people are dissatisfied with the use of maximum sustainable yield as a management objective. There is strong agreement that it is time to get over the expectation that we can simultaneously achieve both sustainability and maximum benefit from every fish stock. This indicates that a conversation about the usefulness of maximum sustainable yield as a biological management objective is long overdue.

In Their Own Words*

*These statements reflect the personal views of interview participants and do not necessarily represent the positions of the institutions or organizations with which they are associated.

Lee G. Anderson, Professor, College of Marine Studies, University of Delaware

Fishery managers are faced with trade-offs. The nature of the problem is that there is no right answer. The public should expect that trade-offs are explicit and dealt with in a clear and forthright way. But neither in the law nor in fishery management plans are managers forced to state what the trade-offs should be. Plan objectives are usually not very specific. Information on the economic effects of management would provide managers with more information on trade-offs to use in decisionmaking.

Neal Coenen, Marine Regional Supervisor, Oregon Department of Fish and Wildlife

Fisheries are in transition. We need to acknowledge that and manage them as such. We've overconfidence in science and the ability to manage. The truth is, marine science is a rudimentary art—it's tougher than working on the moon. We need more humility and to move forward into stable sustainable fisheries. The problem is the intermediate socioeconomic adjustment. Fisheries were built on the expectation of stable growth and that won't be the condition in the future. Constant, long-term yield is a socially and economically desirable objective. Achieving that will require less than optimal yields. We need a discount factor for the uncertainty and risk inherent in fisheries management. We need to bank fish, be cautious, and harvest less fish than we think we could. We need to make the transition from fishing down to stable fisheries.

Rick Lauber, Chair, North Pacific Fishery Management Council

Maximum benefit to the nation does not necessarily mean maximizing economic value. There is a social aspect to benefits as well. The value of fisheries to people in coastal communities, particularly indigenous people, is important and can't easily be calculated. Determining maximum benefit to the nation should not be an adding machine exercise. The Office of Management and Budget approach is not my preference. They might see all factory trawlers as the way to prosecute a fishery based on efficiency. Factory trawlers have a place, but you wouldn't want an entire fishery of factory trawlers.

Mike Nussman, Vice President, American Sportfishing Association

What are we managing these fish for? What are our goals? A stock would be managed entirely differently from a recreational versus a commercial perspective. For the recreational industry, the value of fish is realized only when it's in the water. A fish isn't worth a nickel to us when it's dead. On the other hand, a fish isn't worth a nickel to the commercial industry until it's dead. The recreational industry isn't driven by how quickly they can fish, or how many fish they can take—they don't view fish stocks as crops and they're not concerned with getting the maximum amount of protein out of the water. The fact is, they're not very efficient. They don't know the water or the fish very well. So, in order to have a successful experience, anglers need abundant fish populations and regulations that recognize that they are geographically restricted; that they have to catch what swims by.

Bob Schoning, Former Director, National Marine Fisheries Service

Sustainable fisheries mean keeping stocks and harvests in reasonable balance, recognizing that at times the populations will fluctuate significantly, unrelated to the management regime. In a sense, the fish are not managed, the people are. Nature manages the fish, and the councils try to interpret nature and manage the harvesters to permit reasonable levels of fish to continue. It is important to manage so that the fishing industry is maintained through low abundance years so it will be there during high abundance years. Expecting to maintain consistently very high levels that periodically occur is not realistic in my opinion. More needs to be known about the causes of the wide fluctuations, and regulations should be adjusted accordingly.

Bob Storrs, Vice President,
Unalaska Native Fisherman's Association

Nine thousand years of participation in the fishing industry have been recorded in the community of Unalaska, Alaska. The idea that this can be usurped by well-financed, powerful industry interests is disgusting. There is a lack of recognition that small boats are part of the future. Fishing-dependent communities are more conservation-oriented than the large corporations because communities are concerned with sustaining their participation in the fishery over the long term.

Notes

1. Anon. 1995. *Webster's II New College Dictionary.* Boston: Houghton Mifflin.
2. NMFS. 1996. *Magnuson-Stevens Fishery Conservation and Management Act.* NOAA Technical Memorandum NMFS-F/SPO-23, December.
3. Ibid.
4. Larkin. 1977. An Epitaph for the Concept of Maximum Sustained Yield. *Transactions of the American Fisheries Society* 106(1):1–11.
5. Barber, W. E. 1988. Maximum Sustainable Yield Lives On. *North American Journal of Fisheries Management* 8(2):153-57.
6. NMFS. 1996. *Our Living Oceans. Report on the Status of U.S. Living Marine Resources, 1995.* U.S. Department of Commerce, NOAA Technical Memorandum NMFS-F/SPO-19, February
7. Ibid.
8. Ibid.
9. FAO. 1997. *The State of World Fisheries and Aquaculture 1996.* Food and Agriculture Organization of the United Nations, Rome.
10. Alverson, D. L., M. H. Freeberg, S. A. Murawski, and J. G. Pope. 1994. *A Global Assessment of Fisheries Bycatch and Discards.* FAO Fisheries Technical Paper 339. Food and Agriculture Organization, Rome.
11. NRC. 1999. *Sustaining Marine Fisheries.* National Academy Press, Washington, DC.
12. Personal communication, Clarence Pautzke, executive director, North Pacific Fishery Management Council.
13. NMFS. 1998. *Draft Amendment 1 to the Atlantic Billfish FMP.* Highly Migratory Species Division, 1315 East-West Highway, Silver Spring, MD 20910.
14. NMFS. 1998. *Southeastern United States Shrimp Trawl Bycatch Report.* Report to Congress, National Oceanic and Atmospheric Administration, U.S. Department of Commerce.
15. NRC. 1999. *Sharing the Fish: Toward a National Policy for Individual Fishing Quotas.* National Academy Press, Washington, DC.

16. Ibid.
17. NMFS. 1996. *Our Living Oceans. The Economic Status of U.S. Fisheries.* NOAA Technical Memorandum NMFS-F/SPO-22, December.
18. Ibid.
19. Ecosystem Principles Advisory Panel. 1998. *Ecosystem-Based Fishery Management.* A report to Congress as mandated by the Sustainable Fisheries Act amendments to the Magnuson-Stevens Fishery Conservation and Management Act of 1996.
20. FAO. 1996. A Precautionary Approach to Fisheries, Part I. Guidelines on the Precautionary Approach to Capture Fisheries and Species Introductions. FAO Fisheries Technical Paper 350/1.
21. Randall, A., and M. C. Farmer. 1995. Benefits, Costs, and the Safe Minimum Standard of Conservation. In *Handbook of Environmental Economics*, edited by D. W. Bromley, 26–44. Oxford and Cambridge: Blackwell Books.
22. NRC. 1999, n. 15.
23. NRC. 1996. *Upstream: Salmon and Society in the Pacific Northwest.* National Academy Press, Washington, DC.
24. NMFS. 1998. *Report to Congress: Status of Fisheries of the United States.* U.S. Department of Commerce, NOAA, October.
25. NEFSC. 1996. Report of the 22nd Northeast Regional Stock Assessment Workshop, Stock Assessment Review Committee (SARC), consensus summary of assessments. NOAA/NMFS/NEFSC. NEFSC Ref. Doc. 48-135. Woods Hole, MA.
26. Christy, F. 1997. Economic Waste in Fisheries: Impediments to Change and Conditions for Improvement. In *Global Trends: Fisheries Management,* edited by E. K. Pikitch, D. D. Huppert, and M. P. Sissenwine, 28–39. *Proceedings of the Symposium Global Trends: Fisheries Management,* American Fisheries Society Symposium 20, American Fisheries Society, Bethesda, MD.
27. Fraser, J. 1998. An Industry Snapshot: A Look at Commercial Fishermen Through National Fisherman's Exclusive National Survey. *National Fisherman* 79(8) (December 1998):26–33.

CHAPTER 4

Owning Fishery Resources

The general public has the right to expect that management will consider the public interest.

John A. Mehos, formerly Texas Shrimp Association

All U.S. citizens own resources in the [Exclusive Economic Zone] and they have every right to get benefits from them like they have for land-based resources.

Jim Cook, Western Pacific Fishery Management Council

Fish are a public resource that belong to people in Iowa as well as the fishermen of Gloucester. We need to manage our fisheries for the common good.

Sonja Fordham, Center for Marine Conservation

The American people own the fish in the Exclusive Economic Zone. This is what is meant by "public trust" and "public ownership." But what do these terms imply for managing marine fisheries? Both recognize that governments have jurisdiction over and responsibility for fishery resources. The government acts as a representative of the people—a trustee—as it does for other natural resources. The ideas are also expressed in the public trust doctrine of common law, which strengthens legislation such as the MSFCMA by reminding courts and lawmakers that we the people have ownership rights and obligations with respect to fisheries.

The idea of a public trust is closely connected with two other ideas that remain active, although always contentious, in the United States. First is the idea of free access or the "public right" to fish and wildlife.

Second is the idea that no one can have exclusive, private ownership of fish or wildlife until it is captured. These ideas are so strongly held in some areas that it has been difficult to develop fishery management programs that restrict public rights of access.[1,2] Many fisheries are at a crossroads today as these cherished notions are challenged and reinterpreted.

Expanding Public Interests

We should give the public a robust ecosystem because that's what was there before the trouble began. That's what we need to ensure there are diverse fisheries in the future.

Dorothy Childers, Alaska Marine Conservation Council

The public should be assured that management will lead to a standard where you can catch a fish if you live on the coast, where you can have a seafood dinner if you live in the midland, and, no matter where you are, that we don't wipe out the resource.

Don DeMaria, F/V Misteriosa, *Summerland Key, Florida*

Very often the public interest in marine resources is to allow fishermen, both recreational and commercial, to make money off them.

John H. Dunnigan, Atlantic States Marine Fisheries Commission

Public ownership is a widely recognized theme in fisheries. But, as owners of fishery resources, how can members of the public best realize their interests? The public interest in fisheries has traditionally been based on the idea that those who depend upon and invest their labor and capital in natural resources have priority rights to them. Fishing is seen as a birthright to those who have fished for a long time. Their personal identity with fishing is wrapped up in the idea that "the salt gets in your blood."

The public benefits of an economically healthy fishing industry—including low consumer prices, efficient utilization, and economic contributions to coastal communities—are widely recognized by stakeholders. To some, it is equally important that fishing opportunities be reasonably distributed among user groups. The interests of fishing-dependent communities are also recognized by the 1996 MSFCMA amendments, which require that such communities are taken into account in developing regulations.[3] Although some might think that the representation of fishing industries is adequate for this purpose, there are many members of fishing-dependent communities who have traditionally not been well-represented in management, including crew, processing plant workers, families, and ancillary businesses.

The people we interviewed also recognize the public interest in recreational fisheries, pointing out the benefits they provide to both individual anglers and the coastal communities from which they fish. But balancing commercial and recreational benefits is not easy. Some stakeholders argue that public benefits from recreational fishing go beyond those traditionally considered; for example, appreciation of the environment. "About 60 million Americans know the environment through recreational fishing," one points out. "To them, environmental awareness comes from being out there on that boat." For this reason, another agrees that the public benefits most when fisheries are available to the maximum number of people.

The benefits of recreational and commercial fisheries are widely recognized. But today, ideas about the public interest in U.S. fisheries go even further, illustrated by the suggestion of some that consumers should not only be able to purchase high quality seafood at reasonable prices, but should also be able to do so "without feeling bad." This idea represents an expanded public interest in U.S. fisheries, a change that includes a much wider variety of social interests and activities. "Fishermen and other users of fish are not the total public whose interest we should be trying to serve," a government official remarks. Traditional user groups like commercial and recreational fishermen are now being challenged by the new interests. In light of these new interests, public agencies and professional managers are struggling to broaden their missions and roles.

People who enjoy fishery resources but don't consume them, such as nonfishing recreationists and some contributors to environmental organizations, are now more frequently heard. This group reflects an increasing public concern with environmental quality and biodiversity.[4] While "some people want to have fish available for catch and some want to be able to buy a variety of fish at a reasonable price, others like to see them when they go diving, and many just want to know that their resources are being managed wisely," an environmental representative explains. And, as difficult to accept as the nonuse message has been to some in the fishing industry, many industry members do share its values. One observes: "The public interest of providing quality seafood or providing food in general is obvious but this should be better balanced with maintaining a healthy ecosystem. The council here is just now learning from the public that they place a value on ecosystems and on knowing that things like Steller sea lions are out there. The public should expect healthy ecosystems that remain diverse and should expect a good quality product as well." Members of the conservation community are increasingly able to support this and other perspectives as council advisers, but only one actually holds a council seat. According to one environmentalist, this is "a travesty."

Despite a wider pool of interests, the customary public interest in fisheries—producing seafood commodities, jobs, and incomes—still drives the system in many areas. The traditional fishing identity can lead to powerful rhetoric at management meetings, often making it difficult for alternative perspectives to be heard. As an environmentalist remarks about his experience in the process, "The core industry perspective remains: 'these are our fish and we ought to be able to fish them.' The leaders know this isn't true, but they are mostly grandstanding and playing the crowds at public hearings; . . . they marginalize the new leaders and perspectives."

Broadened public involvement in the fishery management process and rights-based management risk undermining fishermen's views of fisheries, which are based more on cultural tradition. Personal and cultural commitments to marine fisheries are to some extent linked to place. Having cultural ties to fishing—through growing up in a fishing family or living in a community identified with fishing—helps generate the lore, experience, and willingness to take the very large risks that make fishing the productive and adaptable enterprise it has been. Many see cultural ties as generating the personal, familial, and community commitments that support long-term sustainability.

As one industry spokesman notes: "Fisheries should not only supply fish and seafood to the public. Our fisheries provide down-to-earth cultural opportunities for the public to meet and experience the sea. Natural resource communities, including farming communities and fishing communities, have inestimable value to the nation as catalysts for ordering economic activity and cultural education that enable sustainable practices. Our fisheries should involve growth in social and cultural capital as well as economic capital. How else are we going to turn back the tide of overconsumption?" An academic, recognizing the difficulty in assessing such a measure, also emphasizes the cultural benefits of fisheries, indicating that the "net wealth" of the nation includes not just "the sum of economic value of the resource but also the cultural value and human capital."

That the public interest in fisheries means different things to different people is quite clear. In general, commercial and recreational industry representatives recognize ecosystem benefits, but tend to think first of the economic benefits generated by healthy and productive fisheries. Academics also see benefit in "converting fish into dollars." Academics and government stakeholders emphasize the returns to the public that could be generated by fishing in the form of fiscal returns from cost recovery and rent extraction, or cultural returns from maintaining fishing communities and identities. And members of environmental organizations tend to think of the public interest in terms of healthy ecosystems and the aesthetic values of fish—"the pleasure of their company," one calls it.

How best will the public interest be served? A traditional solution is to rely on the continued role of the government in management, because one aspect of this role is to represent public interests otherwise not at the table. The government is often faced with representing an apathetic public.

Public Disinterest

> Fisheries management owes the public factual information, and the councils are the best available way for the public to affect decisions, but the public has often chosen to ignore fisheries management.
>
> Rolland A. Schmitten, formerly National Marine Fisheries Service

Although more public interests in the nation's fisheries are being claimed, the majority of the public do not appear engaged in its rights and responsibilities as owners. For the most part, public involvement in fishery management continues to be limited to representatives of the commercial and recreational industries, academics, and environmental groups. Some stakeholders place blame on the councils for not seeking out the broader public interest. "Look at the councils, with very few exceptions, the public is not welcome," an academic declares. "The League of Conservation Voters, the people who are interested in having fresh halibut in Safeway are not there."

Although important, it is not easy to generate public interest in marine fisheries questions. A fishery manager reflects on this difficulty, saying: "I have come at this in two ways from time to time, and they are not consistent. On the one hand, one could say: What is important about this resource is what I can tell my mother-in-law. As public trustees we have the responsibility to be concerned about it. On the other hand, one can ask: What does she care? It's not real for her, she doesn't eat blue crabs. Let's manage it for the best long-term interest of Chesapeake Bay fishermen and forget about my mother-in-law."

An industry representative expresses similar doubts about the public's interest in fishery management. "I'm not sure if the public interest, of and by itself, truly fits into fishery management," he remarks. "Nowadays, with the political climate, the public may be concerned about the resource but I'm not sure if they're interested in management. Should they be? Yes. Are they concerned with management? I'm not sure. They're concerned with conservation and protection but I'm not sure if they truly know what goes on in fishery management."

An academic argues that the public is not provided appropriate opportunities to reflect their interest in the fishery management system: "I have to be honest and state that I really do not know where the public interest fits into management. We find some signals in the marketplace and other signals via associations. We pick up coastal state input via council appointments. Other than writing Congress or the secretary, we do not appear to offer much opportunity for the inland public sector to provide input to management."

But how much does the public participate directly in the management of any resources? Public concerns are often left to representatives of the public. How far can the government go in managing public resources based on this type of input alone? Can the public owners claim at least some financial return on fishery resources?

Returns to the Public

> If management is costing us more than we're getting from the fishery, then the fishery is generating no wealth to the nation.
>
> Courtland L. Smith, University of Oregon

> The public is being victimized by the inability to capture potential economic rent from the fishery.
>
> Jim Harp, Quinault Indian Nation

Economically productive fisheries offer the potential for more than increased income to the fishing industry and fishing communities. A fishery that is generating healthy profits is a fishery that is in a position to return some of those profits to the public. More and more, people are talking about the level of public subsidy to fisheries in the support of research, management, and enforcement and questioning whether it is in the interest of the public to continue such investment without return. Some have proposed that those who earn incomes from the public's resource also contribute to the maintenance of that resource. These proposals take the form of cost-recovery, where user groups are assessed a fee to pay the cost of management, and rent extraction, where some of the profit generated by the fishery is extracted in the form of a tax or royalty fee. At present, the MSFCMA prohibits the collection of fees over and above administrative costs, except for fisheries managed under individual fishing quotas, where a special fee limited to no more than 3 percent of the value of the individual fishing quota is to be imposed.[5]

Cost recovery is widely accepted as an important change from the historical practice of subsidization. Proponents of cost recovery believe that "at present, the public is being shortchanged." An environmentalist argues, "I don't think that the public should be expected to pay with their tax dollars to bail out people who made poor business decisions." The environmental stakeholders we interviewed argue that industry should be shouldering more of the economic burden of fishery management. Perspectives vary, though, regarding what proportion of the burden should be paid. One recommends that the cost burden be directly proportional to the impact that a particular industry—commercial or recreational—has on the fishery and the benefits it receives. For example, cost-recovery could be achieved through license limitation programs by charging higher permit fees.

According to some, cost-recovery should be applied to all fisheries, rather than just to limited access fisheries. They note that the funds from fees could be supplemented by fines for illegal fishing to help reimburse enforcement costs. At present, these funds go into separate accounts. One stakeholder argues, "This is a public resource and, what is going on right now, it's akin to if we were giving all of the national timber away and we were building the roads and buying the chainsaws." Another thinks that fishermen should cover enforcement costs as well. Some propose that management allow contracts between government and individuals that would include responsibilities to pay for costs of administering and managing the fisheries in return for the privilege of fishing.[6] Contracts could also be developed for groups of rights holders through comanagement arrangements that divide management responsibility between government and user groups. Both mechanisms would promote the idea that exclusive access is a privilege that is "earned and treated as a continual obligation."

A step beyond cost-recovery is rent extraction, in which rights holders are required to return some portion of their profits to the public. It is an idea that is accepted for other public resources such as oil and national forests, where there is a long history of collecting royalties and stumpage fees, but it has never been used in American fisheries. Some think it should be. "When the nation decides to grant privileges to any resource—grazing lands, timber, or fish—the public has a right to demand rent back," states one environmental representative.

Because rent collection is a radical change from historical practice, it is contentious. Stakeholders have strong opinions about whether individuals or community shareholders should pay the public for fishery resources. Many believe that the public interest would be best served by collecting resource rent. Some industry members acknowledge that it would be worth paying some kind of rent over and above cost recovery if they could be assured of a stable and profitable business environment. Many agree that the idea of rents would be more reasonable when tied to some sort of a property right. But the idea of rent extraction during lean times in fisheries is unlikely to be well received, a point made by an observer of the west coast Pacific groundfish fishery, which has recently faced large reductions in quotas. "We need to make the industry more profitable first," he says. Another asserts, "Asking fishermen to pay rent assumes that the landlord can maintain the ocean and its fisheries in reasonable shape to generate profits . . . mismanagement has reduced the stocks to a point where we can no longer talk about rent and things of that nature."

Some argue for rent collection on the basis of making management more efficient, saying, "For efficiency reasons . . . , the amount of revenue should be higher than the costs of management." Moreover, extracting rent can be a way to reduce the cost of entry into the fishery. One stakeholder notes that "if all the rent produced is in the hands of the fishermen, it's impossible for many people without access to capital to enter the fishery." From another's perspective, the development of some mechanism to return resource rent to the public is justified because it benefits all members of society.

Another idea is to collect rents by auctioning off rights to fish. No such system has been used in federal management, in large part because of fear of the radical restructuring of the fisheries that might follow. This fear is well expressed by one industry member: "If you consider a new fishery with no people, auctions [of fishing rights] might make sense. But given the investment levels in existing fisheries, to force head-to-head competition with big deep-pocket corporations where the resource is auctioned off is untenable."

While some indicate support for the collection of rent, others express concern that, if not collected across the board, this idea would attach a "poison pill" to fishery management tools and erode industry support. But even with support for the idea of rent collection, practical realities limit the pace at which these proposals can be implemented. As one fisherman explains, "The public needs to recognize that it has allowed a cult of dependency to evolve in fisheries by the rules that are in place. So while it may be a good thing to shift fisheries off the welfare roles, it will be difficult to do so by wholesale large-scale change." Investments encouraged by subsidies and the open access race for fish are likely to be phased out over time.

In order to assess fees or extract rents, the industry must be healthy enough to be earning profits. And to become profitable, fisheries must move toward management that allows exclusion and places effective limits on fishing. For some this means rights-based fishing; for others, different exclusion arrangements are more appealing.

Private Matters and Public Interests

> Until the structures which control access are resolved, the contestual basis of fisheries will continue and benefits will not be maximized in any generally social sense.
>
> Sam Pooley, National Marine Fisheries Service

> The lack of assured catch privileges leads to the fisheries arms races that wreck fishery management institutions on regional, national, and global scales.
>
> Rod Fujita, Environmental Defense Fund

> The whole idea of recreational fishing is based on public access to the water and fisheries. We live and die by it.
>
> Mike Nussman, American Sportfishing Association

Where does the public trust begin and end? Can the government "trustee" give away rights to public trust fisheries? Can it create exclsive rights or privileges in fishing grounds—including concessions, grants, and leaseholds for aquaculture—and in individual shares of a total allowable catch? Does rights-based management amount to a "give-away" of public resources to private interests? These questions lie behind the debate over public ownership and the private use of fisheries.

The idea of public trust is traceable to Roman law, where it was concerned with access rather than ownership, because the seas and navigable waterways were open to everyone but owned by no one. Fish and wildlife were also "common" property in the sense that they were available to all and could not be owned. During the English medieval period, these ideas changed as nobility claimed fishing and hunting grounds, but these private claims were always in tension with the Magna Carta's protection of public liberties.

The idea of public ownership arose in the sixteenth century when certain resources—the beds and banks of navigable/tidal waterways and the waters and resources in them—were owned by the sovereign. By the eighteenth century, courts had agreed that the king or queen owned this property in a way that recognized and protected certain public rights, namely the rights of fishing and navigation. This is the common law doctrine of public trust, an idea that extended across the Atlantic to the American colonies. Wary of the often dire consequences of the loss of fishing rights, some of the colonies included statements protecting public rights to fish in their charters and constitutions.[7]

After the American Revolution, the people became sovereign and thus the owners of public trust waters and lands. This ownership is exercised by state legislatures, in trust for the people. State ownership includes protection of public use rights of fishing and navigation.[8] The meaning of state ownership has changed over the years, particularly with respect to the states' right to withdraw the property from the public and whether public use rights remain when that is done.

As soon as the public trust doctrine was clarified by the courts in the early nineteenth century, legislatures and other bodies began to lease, grant, and sell underwater plots for shellfish and fish culture as well as sand, oil and gas, mining, and waterfront development. Accordingly, the public trust idea has changed over time, but it remains a legal tool for limiting both government and private action in favor of public rights.[9] The public trust doctrine imposes a constraint on representative government in that legislatures may grant away public lands and waterways, but only if this does not unduly harm public interests in them.

Privatizing fisheries means assigning property rights in the form of catch opportunities to individuals, businesses, or communities. What most people mean by "having property rights" is having exclusive ownership, as well as some freedom to decide how to use the property, including the freedom to sell it. But in practice, property is a "bundle of individual rights" made up of several "sticks." In a fishery the sticks can include the right to catch and enjoy fish, the right of access to a particular fishing ground or fish stock, the right to exclude others, and the right to make management decisions.[10] The composition of the sticks in the bundle varies with different fisheries and different points in time.

In the past, an individual fisherman, whether commercial or recreational, had only one exclusive right—the right to own the fish he caught. Other rights were held in common. As entry to fisheries became limited, those who held permits gained the exclusive right to catch fish. Most recently, rights to some fisheries have been made even more exclusive by assigning individual fishermen the rights to catch a specified share of the total allowable catch.

Creating individual fishing rights that can be bought and sold is an attempt to create benefits by converting most of the sticks in the bundle of rights from shared rights to exclusive rights. Individual transferable quotas, as the most common form of individual rights, are the closest thing to private property in a fishery, where the owner of quota share has exclusive rights to participate in a particular fishery, to take a certain proportion or amount of the fish in the fishery, and, in most instances, to sell or lease those rights to others. These are limited rights, carefully defined in law as permits and privileges whose ultimate ownership remains with the public.

Group ownership of fishing rights by communities, cooperatives, or corporations is also possible. For example, rights can take the form of community development quotas as developed in Alaska, in which a community owns and controls a specified share of the total quota.[11] Rights can also be assigned to a cooperative that then divides its quota share among its members, as established by the American Fisheries Act. Community rights can also be strengthened through participatory and community-based management, often described by the term "comanagement."

Whether individual or community, establishing property rights to marine fisheries is one way to create stable expectations among users and managers. These expectations flow from the rules associated with property. The idea behind defining exclusive rights or privileges is to provide people with the incentives and power needed to develop sustainable fisheries.[12]

Even where rights-based systems allocate resources to specific users, public stewardship interests remain paramount. The government retains the responsibility to conserve fishery resources for the public owners, so fishing rights are more privileges than rights. They can be modified or revoked by the government without compensation. This definition of fishing rights as privileges follows the tradition of Native American tribal definitions under which individual fishermen are given the privilege of fishing, but the right is retained by the tribe.

When fishing privileges are limited, the public gives up the "stick" of free access from its bundle of rights. But it does not necessarily give up the right to have a say in management. When the regional council system began in 1977, the right of the public to participate directly in management decisions for offshore fisheries was added, and so public involvement became easier. An industry stakeholder reflects that "making sure the voice of the public is heard through the council process, represented in the process, and incorporated into decisionmaking is a way to provide a return to the public." This requires care on the part of the agencies and councils to ensure that groups and individuals remain motivated to contribute to the process. The tendency may be for those who receive the first allocation of rights to be virtually the only "public" participants in the process thereafter. As "owners" of exclusive fishing privileges, they may also become the owners, together with the councils and government managers, of the privilege or right of making decisions about how the resource should be managed.[13]

Ideas About Rights-Based Management

[The National Marine Fisheries Service] wants [rights-based management] because it means fewer people to manage. But is that right?

Charlie Bergmann, Lund's Fisheries

[Rights-based management] is the marine equivalent of the closing of the range . . . I embrace the enclosure.

Michael L. Weber, freelance writer and researcher

IFQ [individual fishing quota] is a buzzword that is often used without an understanding of the fundamental change that will be brought about.

Rick Lauber, North Pacific Fishery Management Council

Rights-based management elicits strong views among stakeholders. The overwhelming majority of those we interviewed support such a system and support, at a minimum, limiting access to all fisheries. To one state manager, rights-based systems create the self-interest that fosters stewardship. "If we had more stewardship and cooperation between industry and government," he says, "the government could spend more time and resources in science than in the management process."

Others have strong reservations about rights-based management and are concerned about a number of ways that it could change fisheries. Some fear that fishing rights will be concentrated in a few owners, removing the connection between boats and their customary ports of landing. One argues that the public interest is ill-served when "limited pools of people gain more and more power over access." Others are skeptical about the argument that owning rights to fish will promote stewardship behavior. Some are worried that it will remove fishermen from management and replace them with quota owners, who may never set foot on a boat.

Many who are wary of individual ownership feel more comfortable with the idea of community ownership through programs like the Alaska community development quotas, possibly because they preclude individual concentration of ownership. Still others think that as a matter of fairness, access to fisheries should not be limited to only a few. While recognizing the need to limit access to fisheries, they are strongly opposed to any type of system that allocates rights, including community development quotas.

In recreational fisheries, the approach to the bundle of rights has been to limit access by seasons or areas, or to limit the amount of individual catch through "bag limits," gear restrictions, or requirements that fish caught must be released. For those fishermen, the right not to be excluded from access to fish and fisheries is the biggest stick in the bundle. The right not to be excluded is an expectation of some in commercial fishing as well. This expectation, combined with the public trust doctrine, the tradition of free access to fish, and the idea that fish belong to no one until captured all help make it difficult to develop fishery management plans that assign exclusive rights. Feelings against exclusion are so strong in some cases that when rights-based programs are developed, they "grandfather" in everyone who has been previously involved in the fishery, making it difficult to achieve goals such as reducing effort or capacity. Controversy over exclusion was one factor leading to a four-year moratorium on the development of new individual fishing quota programs in the 1996 MSFCMA amendments.

In contrast to those who strongly resist exclusion, many more stakeholders are worried about continuing open access to fishery resources. There is overwhelming concern about the overfishing and overcapitalization that have resulted from open access and the "race for fish." Many see the result of open access as the loss of potential rents that the public could be obtaining from fishery resources.

Much of the support for rights-based systems is conditional on how such management is designed, how it specifies the exclusiveness of the right, the conditions under which it could be transferred, and the basis for the assignment of the rights. And several note that because management should be specific to a particular fishery context, rights-based management is appropriate for some, but not all, fisheries.

If exclusive fishing privileges are assigned, how can they ensure stewardship or returns to the public? How can the public interest be served? One stakeholder's perspective is that private property allows owners to set limits on exploitation and stay within them, if they themselves recognize the value in doing so. This view of the benefit of private property captures the essence of the argument for why rights-based management systems are expected to lead to better stewardship of fishery resources. A researcher adds about the fishing right: "It can be a useful way of getting to ecosystem management and protection . . . most importantly because it gives holders the power to exclude."

But another researcher's perspective is that the effect of assigning individual fishing privileges will vary depending on the individual and the business: "It doesn't lead directly to better stewardship; it can lead to it, but it depends on the mentality of the people who are participating. I have seen it as part of very bad stewardship for the same reason. . . . The Alaska black cod and halibut fishery appears excellent, but the Atlantic surf clam [individual transferable quota] fishery is an absolute mining operation."An environmental representative agrees: "Property ownership doesn't necessarily produce good management or stewardship behavior. Just look at the variation in home ownership and care."

In discussions of alternative forms of rights to fishery resources, the central question remains: What is in the public interest? All those we spoke to agree that the question of public ownership is central, but there is much disagreement on the question of rights of access and use.

Conclusion

> The benefits the public should expect from its fisheries are high quality, moderate-priced seafood and maximal recreational opportunities. These outcomes will only be realized with hard choices about efficiency and resource allocation among competing user groups.
>
> *James E. Kirkley, Virginia Institute of Marine Science*

> The major issue in my mind is to get straight the issue of access to the resources. Open access to fisheries resources needs to be eliminated totally.
>
> *Michael J. Fogarty, University of Maryland*

Management of land-based resources has been polarized into local versus outsider, wise-use versus public management, and private versus public property debates. These debates have not yet become as extreme in the fisheries, in part because the private versus public issue has been largely the inverse—fishing businesses have been dependent on their public rather than private claims of access to the resources. Accordingly, we have an opportunity to address the question of ownership before it is hopelessly politicized.

As the trustee of fishery resources, the government may assign exclusive rights to the resources only if that is in the public interest, so the question rests on how the public interest is defined. Although some of those we interviewed still believe that all citizens should have access to the fishery resource, most do not support the idea that access should be free and open. Limited access programs have become standard practice in fishery management. Under most of these programs, access is possible but not free. Still, the assignment of rights not only to fishery access but also to a specified share of the catch remains controversial among many as a way to generate public benefit.

In Their Own Words*

*These statements reflect the personal views of interview participants and do not necessarily represent the positions of the institutions or organizations with which they are associated.

Jerry E. Clark, Director of Fisheries, National Fish and Wildlife Foundation

I strongly support some form of rights-based fishery management. Rights-based systems are the best way to take care of the public interest and take us from our present situation of depleted fisheries to a condition of robust fisheries. Parceling out rights will give the recipients something of economic value with which they can make choices. Quota cuts under the present system, and the subsequent loss of investment in boats and gear, just keep putting people out of business. We need to make a national decision on how to transform the system to rights-based fishing. We need to explore the many ways to do that, and that process should begin immediately. I have almost no concerns about the transferability of the rights. My family homesteaded land in the Midwest. At that time, the public gave away land because it was in the nation's best interest.

Robert B. Ditton, Professor of Wildlife and Fisheries Sciences, Texas A&M University

The public interest is served by including politics in the process. For example, if the National Fisheries Institute uses power to affect a decision made by a council, then that decision is in the public interest. We are a democracy. Politics are part of the process in fisheries management so a result that derives from the whole process should be in the public interest. The fisheries management system needs to make better arguments to involve more of the public in the process. We need more effective groups to work together to build a constituency.

David Fraser, Captain, F/V *Muir Milach,* Port Townsend, Washington

Cost recovery is a legitimate public concern. From the industry side there are some preconditions that are necessary before resistance to user fees can be overcome. These relate to the regional application of funds collected, industry–agency comanagement of user fees, and the type of program to which fees are applied. There is a quantum leap between cost recovery and rent extraction. The rent extracted from timber companies is in exchange for secure property rights to harvest a certain quantity of timber. Miners or foresters do not race for the minerals or trees, and then pay rent for what they have extracted. They pay up front for secure access rights.

Bill Gordon, Former Assistant Administrator for Fisheries, National Marine Fisheries Service

In 1972 I argued that some form of limited entry was the only way to achieve sustainable fishing, to transfer the conservation philosophy to the fishing community through economic incentives. Limited entry would have to be packaged so that multiple stocks were included, so that fishermen could diversify over several stocks and adapt to the natural variability in biological production. If we had moved aggressively in that direction we would have received more benefits to the nation from highly profitable fisheries. People haven't debated these issues and kept them before the public long enough for them to become familiar and to make progress. We've never looked really broadly at this. Some of the early opponents of limited entry in the industry are now its strongest supporters. The only way fisheries can survive is to move in this direction.

Jim Harp, Quinault Indian Nation

The tribes have, as a result of the 1850s treaties between the tribes and the federal government, specific, treaty-reserved fishing rights. These treaties specify rights to harvest and gather resources in usual and accustomed places forever. My great-great grandfather was one of the signers of the 1856 treaty between the United States and the Quinault Nation. The tribes' view of fishing rights is that the tribal share is private property that is protected by the U.S. Constitution. Individual tribal fishermen exercise a privilege of fishing but don't have specific rights—the right belongs to the tribe. How to allocate intertribally and within tribes has been the most difficult and contentious issue of the past 20 years. When no one is happy, that appears to be the best allocation scheme. If no one likes it very well it must mean that the pain is being shared.

Margaret F. Hayes, Assistant General Counsel for Fisheries, National Oceanic and Atmospheric Administration

Fisheries do belong to the public, and our management should keep that in mind. The public interest is represented not just by fishermen, consumers, and scuba divers. It is represented by those people who don't fish, eat fish, or even look at fish, even if they aren't aware of fisheries and don't care how they are managed. The government must speak for the disenfranchised.

Ray Hilborn, Professor, School of Fisheries, University of Washington

There are enormous lost social and economic opportunities in present-day fisheries. Our fisheries are managed in the most inefficient way imaginable. Overcapacity and the resulting regulations on communities provide clear examples of this. The resolution of allocation issues would increase the economic efficiency of the fisheries and maintain the potential for social and economic yield and biodiversity. We should stop pretending that we can manage without allocation and should put it on the table to be dealt with explicitly. Allocation schemes should be established for commercial, recreational, and subsistence fishermen. I am highly supportive of rights-based management. If you don't have rights, you don't have management.

Ken Hinman, President, National Coalition for Marine Conservation

I favor privilege-based management. The fish belong to the public and it should stay that way. The public should bestow the privilege of fishing to all its members within agreed upon parameters and revoke that privilege if it is abused. After reviewing the best science available, receiving input from and listening to the advice of the fishery participants as well as the interested public, and considering what is required under the

Magnuson Act, managers have to be willing to stand up to the fishing industry. And the politicians who've written our laws must stand behind them. We need to act based on a precautionary approach so that, if we are wrong, we will have more fish in the water than we expected, not less. If we are wrong, fishing in the future will be better than it is today, not worse.

Rod Moore, Executive Director, West Coast Seafood Processors Association

In reality, we have to have some kind of rights-based management. The question relates to what form a rights-based system should take. This will be different for each fishery and will need tinkering over time. We need to set objectives for a program, look at the industry, figure out how to design a system that works, and recognize up front that some will win and some will lose. It's a difficult thing for many people to accept that the "gummint" can say who can and can't fish. It creates pressure on decisionmakers to try to make everybody winners.

Sam Pooley, Industry Economist, Honolulu Laboratory, Southwest Fisheries Science Center, National Marine Fisheries Service

I think the public would benefit from community-based, contractual management in a broad range of situations. Basically, the idea is that the councils would give a substantial amount of authority to stakeholders for management; including not just commercial fishermen, but the entire community. The government would represent the public in making sure that decisions were ecologically appropriate and would provide standards for conservation and equity of access. The failure of the community as represented by a community cooperative, corporation, or other formal body to meet these standards would result in the decisionmaking power being revoked. A successful contract that promoted resource conservation would be renewed. A failed or ineffectual contract would expire and management rights would return to the government.

Gil Radonski, Former President and Chief Executive Officer, Sport Fishing Institute

Rights-based management is the way to go. Early individual transferable quotas, like the community development quotas in Alaska, ignore the idea of returning to the public something for the resource. The resource was given away. We should sell fish to the highest bidder, just like we do timber. Ownership works in all other renewable natural resources. We don't give away timber out of national forests. But we do give away fish.

David Sampson, Associate Professor of Fisheries, Oregon State University

When I began my career in fisheries science in 1975 I read much of the then new literature about the benefits of limited entry. I thought that limited access to fish stocks made very good sense. I am aghast that we still have not embraced this idea. Why do so many people believe it is better to allow the productivity of our fish stocks to be squandered in an equitable manner than to be limited to a few harvesters?

Lee J. Weddig, Former Executive Vice President, National Fisheries Institute

The reduction of direct users in the fisheries is a major policy issue. In the long term, rights-based management is the way to go. We need to get off the dime on individual transferable quotas. The number of commercial, recreational, and charter fishermen should be restricted just like we govern the number of taverns on a block or taxicabs in a city. The integration of rights-based management into fisheries that are already fully developed is an overwhelming task. We may have gone over the hurdle but there's still the mentality that anyone should have the right to catch a fish if they want. Rights should be leased. Fish are valuable commodities and people would be willing to bid for the right to harvest a certain amount of fish for a fixed time period. This would allow fishermen to amortize equipment and make solid business decisions, at least for the major fisheries.

Richard Young, Captain, F/V *City of Eureka*, Crescent City, California

The big problem in fishery management is matching the harvest capacity of the fleet to the sustainability of the resource so that there is no overcapitalization and excess demand to drive the pressures on the system. You can spend a lot of time responding to brush fires, but if you want to stop brush fires you clear out the brush before the fire season. Ownership is the long-term solution. We have to decide who owns this resource and who has the rights to harvest it, so that people have the incentive to conserve. If the forest were owned and used by everyone, how would we ever have any trees?

Notes

1. McCay, B. J. 1998. *Oyster Wars and the Public Trust: Property Law, and Ecology in New Jersey History.* University of Arizona Press, Tucson, AZ.
2. Johnson, R. D., and G. D. Libecap. 1982 Contracting Problems and Regulation: The Case of the Fishery. *American Economic Review* 72: 1005–22.
3. 16 U.S.C. 1851(a)(8).
4. Kempton, W., J. S. Boster, and J. A. Hartley. 1995. *Environmental Values in American Culture.* MIT Press, Cambridge, MA.
5. NMFS. 1996. *Magnuson-Stevens Fishery Conservation and Management Act.* NOAA Technical Memorandum NMFS-F/SPO-23, December. Section 304 (16 U.S.C. 1854).
6. Rieser, A. 1997. Property Rights and Ecosystem Management in U.S. Fisheries: Contracting for the Commons? *Ecology Law Quarterly* 24(4) 813–32.
7. McCay, B. J. 1998, n. 1.
8. Ibid.
9. Ibid.
10. Rieser, A. 1997, n. 6.
11. NRC. 1999. *The Community Development Quota Program in Alaska and Lessons for the Western Pacific.* National Academy Press, Washington, DC.
12. NRC. 1999. *Sustaining Marine Fisheries.* National Academy Press, Washington, DC.
13. McCay, B. J., R. Apostle, and C. Creed. 1998. ITQs and Community: Reflections from Nova Scotia. *Fisheries* 23(4):20–23.

CHAPTER 5

Managing Fisheries

Democracy is the worst system on earth until you compare it to all others.

Andrew A. Rosenberg, National Marine Fisheries Service

The inertia and drag of the political process means that decisions are incremental and take longer to achieve. Mistakes are big mistakes. The resource may not be able to hold up under the process.

Neal Coenen, Oregon Department of Fish and Wildlife

Until 1977, fishery management was conducted through state agencies and a number of international agreements. Implementation of the FCMA established a new governing structure in which authority is distributed between the federal government and the regions. Congress affirmed in the secretary of commerce the ultimate authority and responsibility for U.S. fisheries, and regional fishery management councils were created to advise the secretary of commerce, develop fishery management plans, recommend regulations, and provide a forum for the public to participate in decisionmaking. Congress believed that involving people with hands-on knowledge and experience in regional fisheries was beneficial and would contribute to the development of effective management plans and regulations.

The act also contains provisions for managing international fisheries. In the case of Atlantic highly migratory species such as tunas and sharks, it allows the National Marine Fisheries Service to act for the United States to manage these species within its own zone in conformity with international agreement. Each international agreement provides its own decisionmaking structure. Some include a role for advisers to the U.S.

delegations to international meetings, in which case the United States conducts public meetings at which interested persons can comment on the U.S. position prior to the international meetings.

In the 23 years since the FCMA was first passed, every Congress but the 103rd—which began the reauthorization process that was to result in the 1996 amendments—has amended or reauthorized the nation's fishery management law. Each change has included major debates over the general direction of management as well as the specific form it should take. Those formal reauthorization and amendment actions are outnumbered by the specific actions taken by Congress on fishery issues, ranging from intervention in the most specific local disputes to concurrence in international fishery agreements.

Who's in Charge?

> There's a basic disconnect between what is spelled out by law and what is exercised. Neither the secretary, nor the regional [National Marine Fisheries Service] director, nor the councils are making good use of the authority hidden in the system.
>
> Burr Heneman, marine policy consultant

The U.S. regional fishery management system involves many players: council members and advisers, public participants, states, tribes, the National Marine Fisheries Service, the secretary of commerce, and Congress. The structure and process of U.S. fishery management are shaped by the law and also by agency policies, council operating procedures, federal rules of administrative procedure, open government and due process, interjurisdictional arrangements, and politics.

The idea behind the regional fishery management councils is to bring regional expertise into management through planning, public participation, and advice to the secretary of commerce. The secretary of commerce has delegated its management authority to the National Marine Fisheries Service, which promulgates regulations based on recommendations from the regional fishery management councils.

Each regional fishery management council has a voting membership that includes the National Marine Fisheries Regional Administrator, directors of state fishery agencies, and public members. Public members are appointed by the secretary of commerce for three-year terms from a nomination list submitted by the governors of the states in a council region. Governors are required to determine that nominees are qualified, which according to law means "knowledgeable regarding the conservation and management, or the commercial or recreational harvest, of the fishery resources of the geographical area concerned,"[1] and to consult with representatives of commercial and recreational fishing interests about potential nominees before they are nominated.[2]

Many tribes retain fishing rights as specified in nineteenth-century treaties with the U.S. government. For example, 26 tribes in the Pacific Northwest hold treaty fishing rights and are represented by a membership position on the Pacific Fishery Management Council. Other councils also have Native American representation, but such representation is required only in the Pacific region. Nonvoting members of councils include the regional or area director of the U.S. Fish and Wildlife Service, the commander of the local U.S. Coast Guard district, a representative from the appropriate interstate marine fisheries commission, and a representative of the U.S. Department of State.

Neither the Federal Advisory Committee Act nor federal conflict-of-interest rules apply to council members, but voting members are required to disclose their financial interests in fishing, processing, or marketing activity that occurs within a council's jurisdiction. New rules in effect under the 1996 MSFCMA amendments require council members to recuse themselves from votes on matters that would have a "significant and predictable effect" on their disclosed financial interests.[3]

Councils must prepare fishery management plans, conduct public hearings, review stock assessments, recommend optimal yields, and design regulations for the fisheries under their jurisdiction. Council members perform these tasks with the help of staff that includes an executive director, administrative support, and professionals such as biologists and economists.

Council meetings—with few exceptions for discussing personnel matters—are public, and timely notice of the date, time, place, and subject of meetings must be provided. Councils are free to devise their own rules and procedures for involving committees and panels in the development and amendment of fishery management plans. Public hearings are held to provide all interested persons an opportunity to be heard in the development of fishery management plans, amendments, and regulations. Public testimony on specific agenda items is submitted in written form and also given orally during council meetings.

In setting up the councils, Congress wanted them to have the additional expertise of independent scientists and interest-group advisors. Councils are required to form a scientific and statistical committee and are encouraged to use advisory panels on a variety of subjects as needed. These committees advise the councils but do not have voting power. Scientific and statistical committees review biological, economic, sociological, and other scientific analyses and advise on a range of scientific issues. Advisory panels counsel on subjects ranging from industry operations, market conditions and the design and timing of regulations to emerging environmental concerns. The eight councils vary in the extent and manner in which they use scientific and interest-group advisors.

States have authority to manage fisheries in their marine waters, which generally extend 3 nautical miles from shore but, from Texas, Puerto Rico, and the Florida gulf coast, extend 3 marine leagues (9 nautical miles). States are represented on the councils and participate in council decisionmaking so that management in state and federal waters is coordinated. The top fishery management official in each coastal state is a designated council member. In cases where federal fishery management plans are absent—for example, Dungeness crab in the Pacific Council region—states have been delegated extended jurisdiction to manage those fisheries in federal waters.

States also involve three interstate marine fisheries commissions— the Atlantic States, Gulf States, and Pacific States Marine Fisheries Commissions—in the management of marine fisheries. These commissions were created to coordinate state regulations and develop management plans for interstate fisheries. The Atlantic Commission has significant authority through the Striped Bass Act and the Atlantic Coastal Fisheries Cooperative Management Act to compel adoption of consistent state regulations to conserve coastwide species.

The secretary of commerce, represented by the National Marine Fisheries Service, has the responsibility to approve, partially approve, or disapprove the councils' fishery management plans and their amendments and to ensure that measures in them are consistent with the national standards and other provisions in the law. If the councils do not meet those requirements, the secretary has a mandate to act. The agency also has the responsibility for developing management plans and regulations for the management of Atlantic highly migratory species.

All rulemaking is in accordance with the Administrative Procedures Act, in the form of "notice and comment" rulemaking. Within five days of receiving a plan or amendment from a council, the National Marine Fisheries Service must publish a notice of its availability. If the agency finds the recommended plan and accompanying rules are consistent with the MSFCMA and other applicable laws, they are published as a proposed rule for a public comment period of 15 to 60 days. After the agency receives and responds to comments, the rule takes effect, generally in about 95 days from the time the council submits a plan.[4]

The National Marine Fisheries Service is now required to submit an annual report to Congress on the status of fish populations and to publish a list of overfished fisheries.[5,6] If the councils fail, within a year of this notification, to develop measures to stop overfishing on these species and rebuild depleted stocks, the secretary of commerce must do so within nine months. The National Marine Fisheries Service has committed to the precautionary approach, both in its strategic plan[7] and in national standard guidelines.[8] Although the guidelines are, by explicit provision in the MSFCMA, not binding, they provide the standard by which the National Marine Fisheries Service and the secretary of commerce judge whether council plans meet the requirements of the law.

The ability of the secretary of commerce to exercise management authority has eroded over time in a variety of ways. Councils have become more powerful in their own right, sometimes enlisting the aid of regional congressional representation in disputes with the National Marine Fisheries Service. Stakeholders increasingly turn to the courts when they disagree with management decisions. These actions have diffused authority and responsibility, giving citizens multiple arenas to obtain accountability in the nation's fisheries.

Congress retains both formal and informal fishery management authority. Formally, it enacts the legislation under which fisheries are governed—including the authorization for the National Marine Fisheries Service. It extends oversight of the National Marine Fisheries Service through the appropriations process and holds oversight power to request studies and reports. Congress calls agency officials to give testimony at hearings and has intervened to overturn fishery management decisions and plans. Informally, Congress extends oversight through inquiries of the National Marine Fisheries Service by members, staff contacts, phone calls, meetings, and other communications with the agency.

Congress has a special role in the allocation of U.S. fishery resources to foreign fleets through approval of Governing International Fishery Agreements, which allow U.S. fishery resources that are not being caught by U.S. vessels to be allocated to other nations. Both the House and the Senate can approve these agreements. In addition, as with any other treaty, the Senate must give its advice and consent to any international fishery agreement negotiated by the United States.

Expectations for Management

> Nobody is perfect for any job, but having knowledgeable people making important decisions on crucial fisheries matters makes sense to me.
>
> *Bob Schoning, formerly National Marine Fisheries Service*

What did the drafters intend when they created this participatory system of fishery management? Did it turn out as they envisioned? The original FCMA envisioned an ideal—a decentralized system based on stakeholder participation, with central authority and federal accountability for decisionmaking. In setting up the system of regional fishery management councils, Congress intended to keep itself out of the day-to-day business of fishery management.[9]

The act called for council members to be "qualified persons," who could make sound judgments in the public interest with respect to the management and conservation of fishery resources. The naming of a federal representative on each council was seen as an important contribution to "a continual dialogue between the councils and the Secretary," in order to avoid serious disputes.[10]

More detail about how council members are expected to represent the public interest is contained in the oath of office:

> I,————, as a duly appointed member of a Regional Fishery Management Council established under the Magnuson-Stevens Fishery Conservation and Management Act, hereby promise to conserve and manage the living marine resources of the United States of America by carrying out the business of the Council for the greatest overall benefit of the Nation. I recognize my responsibility to serve as a knowledgeable and experienced trustee of the Nation's marine fisheries resources, being careful to balance competing private or regional interests, and always aware and protective of the public interest in those resources. I commit myself to uphold the provisions, standards, and requirements of the Magnuson-Stevens Fishery Conservation and Management Act and other applicable law, and shall conduct myself at all times according to the rules of conduct prescribed by the Secretary of Commerce. This oath is freely given and without mental reservation or purpose of evasion.

The legislative history of the FCMA includes debate about the respective roles of the secretary of commerce and what were then the new fishery management councils. Congress anticipated there would, from time to time, be conflicts between the secretary of commerce and the councils. One version of the proposed legislation described a third, judicial branch of the management tripod: a Fishery Management Review Board to resolve disputes. But the board did not make it into the final version of the act. The final authority for rulemaking was clearly placed with the secretary, as was approval of the fishery management plans. At the same time, the debate reveals that Congress meant for the councils to play a significant role, and for fishery management plans to originate with them.[11]

"The regional councils are, in concept, intended to be similar to a legislative branch of government," reads the section-by-section explanation of the Senate bill. "The councils are afforded a reasonable measure of independence and authority and are designed to maintain a close relation with those at the most local level interested in and affected by fisheries management." The legislative history is quite clear that the Congress intended no change in the authority of states to manage within their waters, nor to upset customary international law regarding fisheries.[12]

The legislative history of the act also contains many references to the need for individuals who will represent the "public interest," not any particular vested interest in a fishery. "The original idea was to appoint to the councils long-term fishermen interested in the resources—older stewards with sincere concerns about the future of the fishery," recalls one observer. The notion of decentralized management involving fishery participants was reaffirmed in the most recent reauthorization of the act. Although attempts were made to refine rules about conflicts of interest among council members, it was not suggested during the 1992–96 reauthorization debate that the council system itself be rethought.

One person we interviewed believes that the tension over who is in charge has existed since the enactment of the FCMA, and that, rather than investing in a process to sort out the relationships, "we chose instead to concentrate on the details of substance." The vacuum created by this ambivalence was soon filled. "In the early days, the councils were seen as advisory, and the secretary as the decisionmaker. That has changed so much, that the councils are considered to be in charge," observes one agency official. "In 1976, if the agency had held the line and staked out the position that they are professionals who should be allowed to do their jobs, to act on behalf of the resource, it would have been respected," the official says. "Now, after two decades of getting pushed around, I don't know how you could turn the National Marine Fisheries Service into an independent-minded bureaucracy that could stand up to Congress."

Outcomes

Based on my observations, trivia takes up a tremendous amount of time in the management process.

Timothy Hennessey, University of Rhode Island

> Many times industry members haven't demonstrated a willingness
> to make tough short-term decisions that are isolated from social
> and economic consequences. . . . As members of the community,
> it's difficult for them to divorce themselves from the consequences
> of their actions.
>
> Tom Hill, New England Fishery Management Council

Many stakeholders observe that the council system is an example of
democracy at its best and its worst—open and responsive to constituent pressures but also slow and cumbersome. Some would argue
that, though it's messy, the council process is democratic, and that all
participatory processes are slow. Others see the council as not easily
accessible and call the system an insult to democracy—an appointed
body, made up mostly of people who have power, using their position
to reward others with power, including those who have broken fishery
laws.

An industry member observes that a strength of the council system—the deliberative, consultative process with stakeholders—is also
its weakness. "Years ago we all learned how to throw sand in the gears
to slow anything down," he says. "The council system is better at preventing things from happening than getting things to happen." But this
view is not universally shared. Is it the rule or the exception?

Once management measures are submitted by the councils and reviewed by the National Marine Fisheries Service, the next hurdle is the
rulemaking process. The MSFCMA describes an open, participatory
process at the regional council level, but the business of making rules in
America is serious indeed. Added to the MSFCMA's requirements are the
requirements of several other acts: the National Environmental Policy
Act, which requires analysis of the interactions of the natural and human
environment; the Regulatory Flexibility Act, which requires an assessment of the effect of regulations on small businesses; Executive Order
12866, requiring a Regulatory Impact Review of the costs and benefits of
alternative actions; and the Paperwork Reduction Act, which places
controls over the collection of fishery information from the public.
These rules about rules create a complex minuet of notice, comment,
response, notice, comment, wait, notice, before getting to a final rule.
Although designed to safeguard and equalize public participation,
administrative rulemaking—which is where the rubber of fishery management measures meets the road—is highly structured and very much
an insider's game.

Figure 5.1 illustrates the steps that managers and stakeholders must
go through to develop and implement a fishery management plan. The
120-day process is required by the MSFCMA and administrative rules.

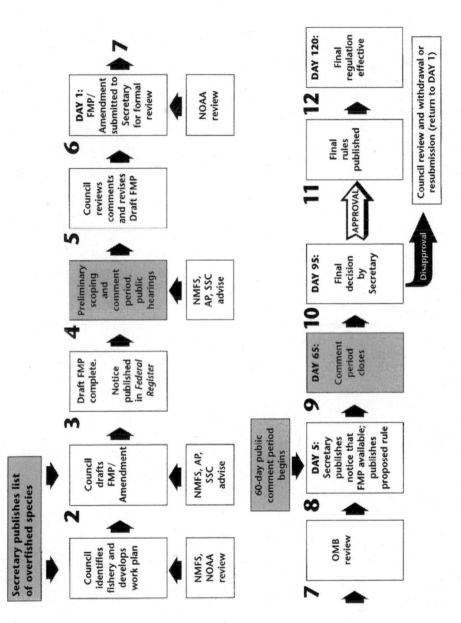

FIGURE 5-1 Schematic showing the various steps in drafting, reviewing, and approving a Fishery Management Plan. Reprinted with permission of the Center for Marine Conservation.

Critics now say that the council process has become overtly political, with candidates for council membership campaigning at every level— for their governor's nomination as well as for the support of interest groups and congressional delegations. The politicization carries over into council deliberations and process, with participants or proponents of a measure driving to gain support and align votes as at a political convention. They are pressed to trade off their votes on one issue in exchange for support on something their interest group may want later. This process disenfranchises individuals, industry sectors, and the public, and one correction would be to constrain the basis for decisions to objective, measurable criteria.

Instead of considering proposed measures with regard to their effect on public resources as a whole and the "greatest benefit to the nation," council members are driven to make compartmentalized decisions that help or hurt one segment of the industry, one gear group, or one port, without regard to the long-term consequences to all participants or the ecosystem. This process provides advantages to certain groups and disenfranchises others. Well-organized associations and gear groups have more clout than individual fishermen, and certainly more than the public at large. Attempts to strengthen conflict of interest provisions governing council members fell short in the last MSFCMA reauthorization. Disclosure and recusal by members who are tempted to vote their interest, designated seats for nonuser stakeholders, reform of the appointments process at both the governors' and the federal level, the role and accountability of state directors as voting members, and taking councils out of scientific and total allowable catch-setting processes should all be examined for their potential to relieve the political pressure on council members.

Some stakeholders feel that the politicization of the fishery management process has eroded the balances between national policy and local concerns, between and among user groups, and between congressional oversight and executive action. As with all political decisions, the "squeaky wheel" gets the grease. The ability of an interest group to mount a political campaign on an issue is often the deciding factor on management outcomes. This can override considerations of science and policy in the national interest and lead to ad hoc decisionmaking and congressional micromanagement. For example, the New England Fishery Management Council proposed measures to restrict groundfish catch in New England in 1990. At the behest of fishermen, the Massachusetts congressional delegation preempted the council process by introducing legislation setting out measures and a time frame for restructuring the fishery. Members of congressional delegations from Gulf of Mexico states have used appropriations bills numerous times to thwart implementation of gear requirements put in fishery management plans to reduce bycatch in shrimp trawl fisheries.

Special Interest or Public Interest?

> As it is now, de facto allocation by the councils is a problem because who gets the fish depends on who is on the council.
>
> *Ray Hilborn, University of Washington*

> Ultimately, political maneuvering caused the problems that we now face.
>
> *Lee G. Anderson, University of Delaware*

The original drafters of the FCMA rejected a notion that council members should represent specific interest groups, fearing that in order to balance the interests, councils would have to be "far larger than would be administratively workable." In the alternative, they called for "qualified persons" and defined a qualified individual as one "who is knowledgeable and capable of making sound judgments in the 'public' [emphasis in original] interest with respect to the management and conservation of fishery resources."[13] They left the balancing act to the governors and the secretary of commerce.

Because nominations are made by governors, they are influenced by highly placed political interests. Final approval authority remains with the secretary of commerce, who has the power to reject any nomination. But one environmental advocate argues that the appointments process should be held up to much closer scrutiny. The outcome is "appointments that mystify" he says, noting that it allows the retention of council members who have been recognized as part of the problem for years.

Because the appointment process is political, council membership then becomes a question of interest group representation, not merit, according to a participant from the recreational sector. A fishery manager shares this view, saying, "The decisionmaking body is comprised largely of partisan representatives of small interest groups who act as representatives of the group rather than knowledgeable individuals acting in the public's best interest." Even though not elected, council members may behave as if they are—beholden to the constituents who lobbied for their appointment. "It's a bad idea to have organizational representatives on the council," according to a west coast participant. "They have to answer to their constituents and can't take the broader view. The appointments process is so highly politicized that what you get are politically well-connected people with no knowledge of the fisheries they are supposed to manage."

Are the public-seat council members typically lacking in knowledge of the fisheries they manage? There were very different views on this among those we interviewed. Some counter the claim that appointed members are uninformed, arguing that many council members—especially those representing various industry groups—have extensive knowledge of the fisheries they manage. Many see council members tending to align with their own interests and vote in blocks. But a government scientist believes this is what has made the councils successful: "If you look at it in terms of how the councils have responded to the perceived needs of people in their regions, they have responded to constituents; they have been represented well." A representative of the sportfishing industry says of interest blocks: "When you do it right, you have a chance, but you need political savvy to do it, a political hammer."

Whether the interest blocks are balanced on councils is also open to debate. Industry members using a particular type of gear think there are not enough council members representing their point of view, the recreational interests think that the commercial sector is overrepresented, and environmental groups believe there are too many user group interests. "It's important to have fishermen involved," says one advocate, "to inject reality into the process and improve input, but it has to be balanced by the views of other stakeholders."

One industry spokesman points out that special interests and conflict of interest were anticipated in the original system: "You'll always have it. The provisions against conflicts of interest in the act now don't mean much. The only control against it is to have a big enough group of people on the councils so you can balance out the interests."

Accountability

> The authors of the FCMA intended to create a process for addressing the public interest but, in reality, they created one for responding to constituent interest. I don't know how to get the average person on the street into the process.
>
> *Rolland A. Schmitten, formerly National Marine Fisheries Service*

> Accountability for management decisions is not there. In the private sector, you must produce results.
>
> *Eugene C. Bricklemyer Jr., Aquatic Resources Conservation Group*

Is it enough to just balance the interests? Or were the councils intended to go further in considering the public interest? What about accountability? And to whom are people accountable? "The councils themselves don't necessarily bear much relation to the public interest," one environmentalist finds. "They are useful for articulating the interests of individual groups, but the public interest is largely absent." Another environmental advocate notes that although there are benefits from involving fishery participants because they're close to the ground, it's often difficult for them "to take the painful steps necessary to put themselves and/or their neighbors out of business."

When the FCMA was written, the placement of a National Marine Fisheries Service representative on each council was seen as an important contribution to "a continual dialogue between the councils and the secretary," and a way to avoid serious disputes. The drafters also anticipated contributions by the regional state fisheries commissions, which had been around since the late 1940s. Today, the federal–state relationship remains an uneasy one. One participant describes the occasion of a National Marine Fisheries Service person sitting through four years of debate on an issue before a council, saying nothing to change the direction of the decision as it evolved, then overturning the decision the council sent to the agency. Some see state managers as even more politically sensitive. As one stakeholder remarks, "When the state fisheries representative is at the table, he's got to be able to deliver, and he knows that means getting it through the state legislature," and all the constituencies that influence it. The fact that the budgets of state fishery agencies are controlled by state legislatures further strengthens the sensitivities of their directors to constituents.

What was envisioned two decades ago as a way to get closer to the people affected by fishery management has often disenfranchised individuals, sectors of the industry, and the public. "In conjunction with the political nature of the appointments process," says one analyst, "they encourage management based on a 'beggar thy neighbor' approach." "Where it used to be us against them, after the foreigners were kicked out, the issue became 'us against us,'" observes another participant. The appointments process and the decisionmaking process had to become more political as members found themselves affecting each other, rather than outsiders.

Without direct accountability, everyone and anyone is to blame for a given decision. It matters little that it is the secretary of commerce who is legally accountable, and who is the respondent in court if that is where a decision is challenged.

Representation

The more you get involved, the more complex it gets. Just look at the Pacific Council—several thousand fishers, four states, 26 tribes, the federal government. It's a complicated system. I've been an interested observer of the process since 1976 and now an active participant. I was a lot more confident that I knew the answers early on than now.

Jim Harp, Quinault Indian Nation

Council regions, although designed to decentralize the decision process into reasonable areas, are seen by some as still too large. Because of the size and diversity of the regions, some feel that council membership cannot possibly reflect the fleets or the fisheries. "The links are lost," claims one state member. "No one fishing feels connected to the decisionmakers, and there's no accountability from the decisionmakers to the fishermen." This is a distinctly different view from that of councils being too responsive to fishermen.

The advisory panels were anticipated to be the primary mechanism through which user groups would be represented. It was here the councils were expected to fairly balance membership among all interest groups, and to select people "to represent those actually engaged in the fishery, and others knowledgeable and interested in the conservation of the fishery. Each council will need to rely on these committees to assist it."[14]

The utility of advisory panels may be one aspect of the councils about which most observers and participants agree. Advisory panels are seen as very useful as both the originators and the sounding board for ideas and proposals that eventually come before a council. Advisory panel members tend to be well-regarded among user groups and conservation advocates alike. But each council has a different policy about how it uses advisers. One criticism is that some councils do not use them enough, and another is that the quality of the selection process varies widely between councils. For example, some councils may not allow anyone with fisheries violations to serve on an advisory panel; others may require that a person serve on only one panel; still others may allow one individual to sit on several panels. Some councils appoint panel membership; others have a more "come one come all" policy. Some pay travel expenses for panel members; others do not.

Checks, Balances, and Monkey Wrenches

The public should really participate in public hearings and meetings. They could bring back into the equation something that we've forgotten; or that the users have forgotten.

Graciela García-Moliner, Caribbean Fishery Management Council

There is a steep learning curve for gaining access to the council
process and high costs for maintaining access to the process.

David Sampson, Oregon State University

The original FCMA envisioned an ideal—decentralized planning arising from stakeholder participation, with central authority and federal
accountability for decisionmaking. A long-time practitioner observes,
with admitted overstatement, that "the act was perfect in 1976. Every
time they've changed it, it has become muddled."

What happened to the ideal? As with any ideal, it was shaped by the
actual. A half dozen or more statutes, hundreds of individual regulations, and thousands of pages of plans, procedures, and guidelines
describe the scope and process by which we decide how to manage
American fisheries. This complexity hints to reasons why the system
does not work better. But even if the standards, rules, and regulations
were perfect, other barriers to effectiveness still exist.

First, participants are not always focused on the public interest. The
politics of scarcity can sometimes place a premium on self-interest, leaving no room for the public interest. Some see a very important need to
get people more involved in the fishery management process, as well as
a simultaneous need to determine how wide a scope of interests should
be included. Increased participation means a corresponding need for
investments in education and training. "The root of the problem is public ignorance about how to get involved," says a manager from New England. He cites the process as intimidating for fishermen who do come to
meetings and "are bombarded with management decisions based on scientific terms like F_{MAX} and $F_{0.1}$. They don't know what's going on, what
the council is talking about, or how to respond."

Council participation can also be kept diverse or the diversity
increased by designating seats, by strengthening procedures for conflict
resolution, or by getting politics out of the appointments process.
There have been efforts by institutions such as Sea Grant and fishermen's associations to provide education and training in the council and
other management processes. Conservation advocates, too, have been
training and urging their members and supporters to attend council
meetings and participate in the deliberations, as well as make views
known through letters and other action devices.

Second, many of those we interviewed say that council members are
sometimes ill-prepared for the job. Or maybe it's that the job requires
more than can be expected from people. "It really requires that there be
a pool of talented people who have the respect of their peers, are knowledgeable, and can work with policy and management people, but have
credibility as fishermen," observes an analyst from New England. "I'm

afraid we haven't had many people like that, or at least many who were willing to put in the long, frustrating hours or who had the political connections to get themselves appointed." A west coast writer and analyst says it would be difficult to ask anyone to serve on a council, given the little support he or she would receive. "They are dealing with very complex issues, sophisticated analyses, and no one should be expected to be born with that understanding." And a former council member recalls "the utter illiteracy of the industry people toward science," charging that "they did not have the background and the training, could not read a scientific paper, did not comprehend the basis of sampling."

Under the present system, new council members are offered an opportunity to attend an orientation or training session to learn the rudiments of the management process, but participation is not required. Some stakeholders who have attended these sessions praise their utility; others claim it would have been better to plunge right into council work and attend the training after a year, by which time some of the terminology and concepts would have made more sense. Considering the importance and breadth of their responsibilities, many feel that, at a minimum, some training should be required for all council members.

A third reason that some councils have been less successful than anticipated is that whenever fish stocks have diminished, long-term planning has taken a back seat to immediate needs. It is small wonder that there is no room for strategic considerations, review of fundamental principles, or exploration of long-term goals. And in fact, the law does not specifically require that councils do long-term planning. The work required of the councils focuses on the short-term tasks of annual decisionmaking.

One manager asserts that the councils and the National Marine Fisheries Service never worked out their respective responsibilities and authority. The system is limited in fostering accountability for those decisions because it is unclear who is deciding. The only way you can have accountability, according to one analyst, is to know that there will be decisions made with which constituents disagree. "But you can't do that in a representative democracy without loading up the congressional mailbox." Even if one accepts that decisions are made in the executive branch, by the agency not the councils, some point out that "the Department of Commerce and the National Marine Fisheries Service are extraordinarily sensitive to the same constituents. The service does not take the stance that good management requires hard decisions, which cause reactions they are willing to face. They are scared of every elected official, especially appropriations committee members."

The process has had still more layers grafted upon it in the form of the 1996 amendments.[15] The MSFCMA limits council authority by forcing them to meet specific standards, leaving them far less discretion. But constituents unhappy with the outcome still have other avenues of influence, and that's part of the problem.

Oversight is supposed to occur at the National Marine Fisheries Service, to ensure that council plans and recommendations meet the national standards and other requirements. The agency points to more plans disapproved by the secretary of commerce in the past several years than at any time in the past. But council participants claim that the agency has done a sloppy job watching over council decisions or stepping in when a council has done poorly. "They're afraid to step in and they lack the firepower to step in," one participant claims.

Several say there needs to be a clearer separation between conservation and allocation decisions—a separation between setting the total allowable catch and deciding who gets the catch, just as the 1986 Blue Ribbon Panel recommended. Many advocates, scientists, managers, and observers suggest that such a separation would take many contentious conflicts out of the council process. With this approach, setting total allowable catch would be done by government and academic scientists, who would then present a firm catch limit to the councils where fishermen, academics, conservationists, and others could debate plans for how to stay within that catch and allocate among various user groups.

Several stakeholders also see a critical need to develop clear criteria to guide who gets the fish. To a few, this would include specifying allocation criteria in the MSFCMA, particularly for allocation between commercial and recreational fishermen, although others would strongly prefer that this be left to the regional councils. Some are concerned that recreational fisheries don't fare well in either allocations or regulations.

A long-time practitioner observes that over the years the agency has let the councils take control, rather than giving advice or stepping in when firm action was required: "There have been instances where the National Marine Fisheries Service knew the measures submitted by councils were not good enough, but because of congressional or constituent pressure, let them get away with less than what was required."

Congress plays an oversight role in fishery management, but at what level is it exercised? Critics argue that the legislative body has been micromanaging through regulatory intervention, special legislation, line item appropriations, and budget language. Examples of ad hoc policies that have resulted from this type of congressional action include delays in the implementation of bycatch reduction in the Gulf of Mexico, intervention in the setting of total allowable catches and adoption of management measures in New England, fleet buy-outs in New England and the Pacific Northwest, a moratorium on the development of new quota share programs by councils, a loan program for the purchase of individual fishing quotas in Alaska, a special catch cooperative in Alaska, and the collection of fees in Alaska as an exception to the MSFCMA provision that does not allow any other council or the secretary of commerce to collect user fees.

The fundamental problem with such measures is that they result from political deal-making rather than sound management principles. Often they establish national policy in response to local problems and claims. Such policies can have unintended consequences when they are institutionalized and generalized beyond the special circumstances that brought them into being. Says one council member, "Now anyone who is dissatisfied with council decisions goes to Congress," with the result that some policies are dictated for all council regions on the basis of what's good for a single state. The moratorium on the development of new individual fishing quota programs is an example often cited in our interviews.

Oversight also occurs in the courts, and many participants are concerned that litigation will play an increasing role in how decisions are made. One manager observes that people will use any and every system available to them to challenge the decisions made by those in charge.

It is no wonder the councils can't look at the big picture, let alone develop strategies and tactics for comprehensive management. "Councils are not the place to come up with new ideas; they are locked into their agendas," remarks one observer. "Right now they are like land use planning boards during a development boom—with no chance to plan for growth, but having to deal with permit applications all the time." If councils are to have the space and time to do any long-term thinking to set fishery goals, move toward ecosystem management, or protect essential fish habitat, they need a different atmosphere than the pressure cooker of allocation, deadlines, disputes, and threatened litigation.

Conclusion

> The problem is that the house is burning down and it's difficult to sit still and make rational decisions. We're in crisis management.
>
> Mike Nussman, American Sportfishing Association

It is clear that there is confusion and uncertainty about the various jurisdictions and authorities in management, and that these need to be clarified. Education is needed about who has control over which areas and why. Are we willing to take steps to clarify the respective roles of the councils and the National Marine Fisheries Service? Also needed is a national strategy for fishery management that leads to the development of specific regional plans.

Whatever the respective authority between them, the councils and agency managers should be entitled to an expectation that lawmakers will stand behind those who implement the laws, rather than intervene on behalf of constituents.

In Their Own Words*

*These statements reflect the personal views of interview participants and do not necessarily represent the positions of the institutions or organizations with which they are associated.

Robin Alden, Former Commissioner, Maine Department of Marine Resources

I still believe in the theory behind the regional fishery management councils. But they don't get down to a local enough scale. The outcome then is power politics. There are 17,000 fishermen in the state of Maine represented by only three public members. There's no way those three people can know what they need to know in order to make adequate decisions. The knowledge around the table has often helped the National Marine Fisheries Service and the council make better decisions, but there's a fundamental difference between giving people advisory status and giving people responsibility for management.

Dayton L. Alverson, Chair, Natural Resources Consultants, Inc.

My general viewpoint is that fishery management failure is wedded in institutional paralysis and political involvement. The regional fishery management council process is weak in that it's open to considerable influence by particular special interest groups—not just commercial fishermen, but recreational and advocacy groups too—so there's a lot of inherent special interest that has been guarded very well. Social paralysis resulting from both institutional arrangements and the inability to enforce decisions leads to delays in decisionmaking.

Ray Bogan, Legal Counsel, United Boatmen of New Jersey and New York

Recreational fishermen of low socioeconomic status are not represented in the fishery management process. The management system has tended to disregard their needs and the trend is against them in terms of more regulations and restrictions. Little weight is given to the impact regulations will have on people of low socioeconomic status. Those fishermen usually want to bring home as many fish as they can. What they can bring home tends to be their gauge of success and helps determine whether or not they can justify another fishing trip. What they don't use themselves, they share with friends, usually in the same economic position. They're one of the primary reasons that the ability to take fish home to eat needs to be clearly articulated in the law. Recreational fishing is presently defined as fishing for sport and pleasure—we must add "fishing for food" to this definition. We must also include small recreational communities in the definition of fishing communities.

William T. Burke, Professor of Law, University of Washington

Political dominance and interference weaken the council system. Fishery management is a heavily political affair. This is influenced by how governance of the system is organized. Congress should keep its hands off and let managers do the job according to the science and economics that they know. Senator Magnuson is reputed to have said he would not interfere once the system was in place. Yet congressional meddling wrecked the fishery on the East Coast. What can you do if the very people who created the system sabotage it and won't leave it alone? The biggest single weakness is the ease with which politicians interfere at the request of their constituents. This is why we need to educate the constituency.

William J. Chandler, Vice President for Conservation Policy, National Parks Conservation Association

I have no objection to involving people experienced in fisheries and allowing them to advise the process. But we've got to have managers who will rely on science to make the tough decisions related to setting take levels. Economic interests are always challenging the data, and fisheries managers have too often erred on the side of incaution rather than prudence. The pressure is on the councils to bend this way and that, and there is not enough pressure in the other direction. I'd much rather have a wise judge making the final call than rely on voting by commercial interests.

Jim Cook, Chair, Western Pacific Fishery Management Council

Through the years, the council process and issues to be addressed have become so complex that it's almost impossible for part-time council members to generate results. The system has become too demanding and unwieldy. The complexity probably resulted from the public's insatiable demand to have more people included in the council process, to have more aspects considered in the decisionmaking process, and to have more factual information included in decisions. There are too many aspects and interests that management has to try to address simultaneously. The reauthorized Magnuson Act reflects the most recent event in the evolution toward more complexity. The councils and the National Marine Fisheries Service view the act as a burden more often than not. There needs to be a clear definition of both the councils' and the National Marine Fisheries Service's duties.

Michael J. Fogarty, Associate Professor, Chesapeake Biological Laboratory, University of Maryland Center for Environmental Science

Development and evaluation of fishery management options should be based on participation by all stakeholder groups. The fishing industry and other interested groups should be encouraged to make recommendations and to lend their expertise to the process of formulating management advice in consultation with scientific advisers. I believe that the final management decisions should be made by regional boards composed of individuals with no vested interest in the outcome of the decisions. The board would have standing advisory groups representing industry, the environment, and consumer interests. The board would be obligated to listen carefully to the advice of these groups but would hold the responsibility of making final decisions based on the best available scientific advice and the recommendations of the advisers.

John Green, Former Commissioner, Texas Parks and Wildlife Department

There needs to be very careful consideration of who is appointed to the councils. We need a system of checks and balances to ensure that recreational and commercial fishermen are on an equal footing. Recreational fishermen don't mind limits. It's important to know what amount of fish should be left unharvested to maintain the stocks. But commercial restrictions should be greater than those on recreational fishermen because commercial fishermen are more efficient at harvesting fish.

Jim Kendall, Executive Director, New Bedford Seafood Coalition

I truly think that, with interest and value in itself, fisheries management should be within its own government department. Fisheries should be taken out of the Commerce Department and a Department of Oceans should be created. This would give fisheries and the oceans greater attention. By creating a secretarial position and a Department of Oceans we could look at the ocean as its own entity and decide what is needed. We could organize it to fit its own role. What better way to begin with a clean slate? The time has really come for this. The ocean is not a second-hand stepchild. It's a full member of the family and should be treated as such.

John A. Mehos, Former President, Texas Shrimp Association

The council system established by the FCMA is important in involving the public. It allows for the involvement of all interests in the management process. The FCMA forced people to recognize competing interests, views, needs, and approaches to issues, such as sport versus commercial, big versus small boats, foreign versus domestic fishermen and vessels, and water impacts versus land uses. It provided a mechanism to identify competing interests and to recognize that each has to participate in dealing with trade-offs together. Fishery management should benefit all people involved in and affected by a fishery, from fishermen to consuming public. The council system is the only thing that we have to combine disparate users and interested parties to generate decisions in a collaborative manner.

Rolland A. Schmitten, Former Assistant Administrator for Fisheries, National Marine Fisheries Service

Congressional intervention in the fishery management process often results in lost opportunities. The moratorium on individual fishing quotas is causing lost opportunities because the full array of limited access tools is not available to fishery managers. It may take 10 years to fully use IFQs as a tool as a result of congressional action in 1996. In addition, congressional direction of the National Marine Fisheries Service's budget creates lost opportunities by trying to satisfy regional needs at the expense of national needs. Science has been the biggest loser in that. For example, the earmarking of funds in the Resource Information Line for nonmarine-related activities contributes to many fish stocks' status remaining unknown.

Peter Shelley, Vice President, Conservation Law Foundation

How has the United States exercised its sovereign governance—its exclusive economic control—of this large area of marine resources? There is still no real vision or strategy endorsed by Congress and the White House for managing these resources. It's scandalous for the nation not to have a clearly defined policy for how this commonly owned property ties into our national interest and how it ought to be managed in terms of development, protection, and research. It's like laying claim to all of the Louisiana Purchase and saying "get at it folks, get out of it what you can, and we'll come along later and clean up." Fisheries management is constituency driven, without respect to whether actions further national objectives. Congress is operating under the misguided assumption that insiders know best.

Roger Thomas, President, Golden Gate Fishermen's Association

It's difficult for the councils to apply regulations on an equitable basis when there's a lack of information and when habitat issues, like water supplies, are not under the control of the councils. Water allocation and habitat are the key issues for salmon. Lack of water for salmon hurts fishermen both directly in lack of fish and indirectly in additional regulations. Yet inland water use is not under the control of the council. We need a system in which the whole regulatory package is included.

Notes

1. NMFS. 1996. *Magnuson-Stevens Fishery Conservation and Management Act.* NOAA Technical Memorandum NMFS-F/SPO-23, December. Section 302 (2)(a), (16 U.S.C.1852).
2. Ibid. Section 302 (3).
3. NOAA. 1997. *A Guide to the Sustainable Fisheries Act, Public Law 104-297.* NOAA Office of General Counsel, February 1997, U.S. Department of Commerce, Washington, DC.
4. *Administrative Procedure Act.* 5 U.S.C. 551 et seq. (1946).
5. NMFS. 1998. *Report to Congress: Status of Fisheries of the United States.* U.S. Department of Commerce, NOAA, October.
6. Notice of Overfished Fisheries. 63 *Federal Register* (January 5, 1998): 206.
7. NMFS. 1997. *NOAA Fisheries Strategic Plan.* U.S. Department of Commerce.
8. NMFS. 1998. Final Rule National Standard Guidelines. 63 *Federal Register* (May 5, 1998) 24212.
9. Anon. 1976. *A Legislative History of the Fishery Conservation and Management Act of 1976.* Committee printing, 94th Cong., 2d sess., October 1976. U.S. Government Printing Office, Washington, DC.
10. Ibid.
11. Ibid.
12. Ibid.
13. 1976. *A Legislative History of the Fishery Conservation and Management Act of 1976, Together with a Section-by-Section Index, Prepared at the Request of Hon. Warren G. Magnuson, Chairman, Committee on Commerce, and Hon. Ernest F. Hollings, Chairman, National Ocean Policy Study, for the Use of the Committee on Commerce and National Ocean Policy Study.* October. U.S. Government Printing Office, Washington, DC.
14. Ibid.
15. NOAA. 1997, n. 3.

CHAPTER 6

Creating Incentives

We need to make economic signals consistent with our long-term vision for U.S. fisheries.

Scott Burns, World Wildlife Fund

A lot goes back to the guy on the deck. He doesn't get paid unless he catches fish.

Bill Gordon, formerly National Marine Fisheries Service

However done, the basic idea of defining benefits that would accrue through conservative behavior should be a key idea in fishery management.

Rod Fujita, Environmental Defense Fund

Post-war U.S. fishery policy provided incentives to accomplish the goals of rehabilitating, domesticating, and expanding fisheries. Fishing territory was extended, the construction of vessels encouraged, and fisheries products promoted.[1] But the encouragement of fishery expansion also created incentives to take risks and overuse fisheries. Fish were caught in a race that all could enter, and delays in limiting access to fisheries encouraged investment in fishing capital to compete in that race. This investment resulted in more fishing capacity than the resource could support. To contain the effects of overcapacity, complicated fishery regulations were developed to make fishing operations inefficient. These regulations boxed people into narrow specialization, limiting their flexibility and encouraging in some regions the sequential fishing down of fish stocks.[2] Adding to the complexity, recreational fisheries also expanded, particularly in "sunbelt" areas with increasing retirement populations.

Incentives to grow were well suited to the view of American fisheries as frontiers, in which new fisheries waited as opportunities to be

developed.[3] But the frontier is gone; there are few, if any, fish stocks yet
to be discovered. Management now requires different kinds of behavior. Development is needed less, stability more. Innovation is needed
not for ways to catch more fish, but for ways to reduce bycatch, rebuild
stocks, reduce fleet size, resolve conflicts, and increase accountability.
Management must incorporate multiple objectives, make cost-effective
decisions, and broaden its scope from single species to ecosystems.[4]
Incentives are needed to promote not expansion, but conservation,
stewardship, and stability.

What are our long-term objectives for American fisheries? What signals do our current policies send, and how do they depart from the
ideal? What is needed to realign these signals with the long-term
objectives?

The Ideal and the Actual

> People blame the industry, but the industry is responding to the
> incentives that have been provided. To change behavior, you have
> to change the framework. . . . We need some way to focus on long-
> term changes where people have an incentive to conserve. . . .
> We've got a bunch of four-year-olds in the sandbox and there's no
> adult there.
>
> *Richard Young, F/V* City of Eureka, *Crescent City, California*

> Management is necessary, but so is the cooperation of users.
>
> *Graciela Garcia-Moliner, Caribbean Fishery Management Council*

The MSFCMA and its 10 national standards (listed in chapter 1)[5] set
out the large goals and objectives for fishery management. In their most
general form, these goals are to conserve and protect the biological
integrity of fishery resources, to maintain the economic health of fisheries and fishing communities, and to promote stewardship through
participatory decisionmaking based on scientific information. The
goals and objectives are ideals that express expectations about using,
managing, and sustaining marine fisheries.

Fishery management is expected to maintain fisheries in biological
and economic health. This requires taking action to protect the biological
productivity of stocks, promote the economic productivity of commercial and recreational fleets, safeguard recreational fishing opportunities,
and balance capital investment with the fishery resource.

Fishery managers—agency staffs and council members—are expected
to make responsible decisions. This requires specifying measurable

objectives, systematically assessing alternatives in terms of their ability to meet those objectives, using the best scientific information available, communicating with and educating participants, evaluating management performance, learning from experience, and being accountable for decisions.

Fishery participants are expected to participate responsibly, by learning how the system works, contributing to the design of enforceable regulations, fishing cleanly, and complying with regulations.

Both fishery managers and fishery participants are expected to ensure the stewardship of fishery resources. This involves strategic planning for desired outcomes, tying rights to responsibilities, enforcing rules fairly, and securing tenure in the fishery so that stability and predictability are enhanced.

Are expectations being met? In some cases. But overall, according to those we interviewed, reality is far from ideal. Incentives have a role to play in bringing the two closer together.

Maintain Healthy Fisheries

> The main thing is to reduce or eliminate disincentives, incentives that produce irrational outcomes.
>
> *Michael L. Weber, freelance writer and researcher*

> We need limited entry where fishermen feel they have something to gain by protecting the resource. In open access everybody is grabbing as much as they can. You're a fool if you don't, too.
>
> *Don DeMaria, F/V* Misteriosa, *Summerland Key, Florida*

The fishing community widely recognizes that fisheries are in transition. To many, the MSFCMA is seen as a good change, with protective provisions that should eventually lead to healthier fisheries. But are the incentives aligned so that in implementing the act the management system contributes to long-term goals? As one state manager notes, since the 1996 amendments, councils and the National Marine Fisheries Service regional offices haven't stepped back to assess how the transition can best be accomplished "because they have been on a treadmill trying to meet the new regulations."

Some worry about continuing threats to healthy fisheries from overcapacity, overfishing, bycatch, and ineffective participation in fishery management. Overcapacity is the legacy of past practices such as open access and subsidies and is the fishery condition that most immediately threatens the ability to restore fishery health.[6,7] As one industry member puts it, "The incentives are for the fleet to expand during the fishing

down process. There are high catches per unit effort, lots of money—
the signals say to people 'get in.' By the time we are done fishing down
[virgin stocks], the fleet is much larger than the sustainable yield of the
fishery." Overcapacity creates other problems such as bycatch, pressures to overfish, allocation conflicts, and eventual loss of profitability.
Overfishing and bycatch—the most commonly cited fishery problems—are outcomes of overcapitalization.[8]

How to reduce fishing capacity in commercial, and even recreational,
fisheries is on many stakeholders' minds. "If there is a significant reduction in effort, there will be a lot fewer participants and the political
nature of the process will change," observes a fishery manager. Most
important is to change the incentive that led to overcapitalization—the
competitive race for fish that happens when rights to fish are not
assigned. The race can be removed by regulations like trip limits that
control the amount and frequency of individual landings or the share-
based allocations that focus on making the most efficient use of a share
and on maintaining the health of the fishery as a way to protect the size
and value of that share.

Capacity can also be reduced, one stakeholder notes, through the
reduction or removal of some of the direct and indirect subsidies that
fisheries receive. Examples of direct subsidies to commercial fisheries
include the 1970 Capital Construction Fund that defers payment of
federal income taxes on money used to construct or refit a vessel and
the 1973 Fishing Vessel Obligation Guarantee Program that provides
low-cost loans for the construction and refitting of vessels and shoreside facilities.[9] Indirect subsidies include various research programs
directed at assisting development of the fishing industry.[10] Recreational
fisheries also receive subsidies through the Federal Aid in Sport Fishery
Restoration Act (the Wallop-Breaux Program) in which excise taxes on
recreational fishing equipment, fuel, and other fishing-related goods are
paid into a trust fund that supports state recreational fishing programs.[11] The costs of the program are mostly offset by its revenues, so
it is to a large extent self-supported, but it nevertheless represents a
subsidy to the fisheries.

Other stakeholders worry that the commercial fishing industry will
be at an unfair disadvantage internationally if subsidy programs are
removed. They point out that some of the subsidy programs—for example the Fishing Vessel Obligation Guarantee Program—were put in
place as a balance against subsidies to other sectors that raised prices to
American fishermen. The Jones Act, requiring vessels fishing in U.S.
waters to be built in U.S. shipyards, increased the cost of fishing vessel
construction.[12] In addition, industry members note, American fisheries

are not independent of fisheries worldwide because they compete in a global market where subsidies to fishing are widespread.

The environmental representatives we interviewed tend to focus on incentives to rebuild stocks, particularly in multistock fisheries. There is concern that current law may allow overfishing of some stocks in a mixed-stock fishery. The stocks that are most in need of rebuilding may remain at low levels. But at the same time many recognize that there are practical difficulties to requiring that no species be fished for levels of catch greater than the maximum that can be sustained over time. In mature mixed-stock fisheries, requiring all stocks to be maintained at maximum levels could shut down entire fisheries during a long rebuilding period, creating very high economic costs in lost catch opportunities. A government scientist suggests that targeted taxes—such as landings taxes on heavily fished species—could be used as a way to create economic incentives to reduce fishing pressure on overfished stocks and encourage fishing on healthy stocks.

Bycatch—and the waste of fish it represents—is frustrating to all fishery participants. Even in catch-and-release recreational fisheries, some fish die. Trying to find legal solutions and provide incentives that encourage cleaner gear and less impact on the ecosystem as a whole should be built into management decisions, suggests one council member. Another person we interviewed suggests the need for greater legal flexibility to experiment with incentives to reduce bycatch. One of the alternatives is to land all fish caught, including bycatch, and sell it for public purposes. Other alternatives might assign individual bycatch quotas or reward demonstrated improvements in lowering bycatch rates. Says another stakeholder about Pacific halibut bycatch: "The loss of halibut through discarding is crazy. There should be a program that would allow retention of bycatch for sale, with the proceeds going for some fishery management purpose." In the North Pacific, salmon and halibut bycatch are allowed to be landed and donated to food banks, but not sold.

Make Responsible Decisions

There needs to be a perception by resource users and others that there is a system in place capable of detecting those who chose to operate outside of the rules.

Vince O'Shea, U.S. Coast Guard

If there was ever something overlooked by managers and scientists, it's the actual world of commercial fishing and the industry response to management.

James E. Kirkley, Virginia Institute of Marine Science

Many of those we interviewed feel that the councils and the National Marine Fisheries Service need to be better at making decisions. Stakeholders we interviewed said that decisions are reactive, change frequently, are made with little accountability, are uninformed, are obstructed by legal barriers, and are clouded by conflict of interest.

Reactive decisions have increased as fisheries have become more heavily exploited and the number of problems has increased. The regional fishery management councils have very full workloads. "There are too many demands on people in the short term that prevent the careful thought necessary to capitalize on their skills," says one manager. "Councils have become overburdened," says another. This, an academic observes, causes the councils to be reactive, rather than forward thinking.

One way to fix the problem of reactive, short-term decisions is for councils to be more specific in their goals and objectives for a fishery, and to make decisions in reference to them. Vague objectives in fishery management plans that allow almost any action to be taken are cited as a big problem in forward planning. While recognizing that there needs to be public input into the formation of management objectives for a fishery, some see a need to ensure that objectives are clear and measurable. "We need to identify a goal for management decisions and then build incentive programs around reaching that goal," says an enviromental representative. "We need clear objectives. Let the National Marine Fisheries Service get to the management objectives without a lot of public torture. Then require accountability for the performance of management to those objectives."

Decisionmaking could also be less reactive if councils had more flexibility to experiment. A government official points out that by more clearly specifying objectives, councils could actually buy themselves more flexibility in planning. If councils define standards with a factual basis, come up with ideas that work, and build a record, many legal obstacles to experimentation would be removed. Although it seems obvious that management would begin with a goal, then build incentive programs to reach that goal, the point is made repeatedly in our interviews that this has not been done in an effective way.

Because council decisions are often reactive rather than proactive, they can be too late to stop the consequences of inaction. Default management measures built into the law might help. One manager argues that if thresholds were defined and linked to automatic management responses, fishermen would have more of an incentive to anticipate crises and employ their creativity to develop alternatives. An important change, proposes one stakeholder, would be to ensure that management councils and the National Marine Fisheries Service develop long-run policies that reflect industry practices and provide industry the stability to plan for the future. One way to reflect industry operations, one notes, is to allow individuals the flexibility to diversify over resources and gear types, a factor that would have to be explicitly accounted for in the design of limited access programs.

The number and rate of changes in regulations have increased over time, and some industry members feel the cost of these changes in their inability to plan a stable business. Inconsistencies in management approach erode industry's ability to contribute to problem solving: "We lose the industry's ability to be innovative because they recognize that solutions offered may not stick in the future," says one industry member. "Long-term consistency in the council process would help a lot." An academic supports this point, stressing that instability in regulations creates the incentive to overfish. "Regulations change so frequently that industry becomes skeptical about management and is forced to earn as much as possible at any point in time." When management is unstable, industry may be concerned about the future but be unable to plan for it.

A big concern with many is the lack of accountability in decisionmaking. "Everybody can blame somebody else," a researcher says. Noting that blame is often directed at industry when fish stocks are over-fished, several industry stakeholders suggest that greater accountability should also be expected of the councils and of the National Marine Fisheries Service. One New Englander remarks that the council is "almost an insult to democracy: an appointed body, mostly of people who have power—and they are not always the people who should be making decisions about the fishery. . . . Should Microsoft be crafting antitrust legislation?"

What incentives would promote accountable behavior by council members? Some see council composition as part of the problem. Deciding on fishery objectives, total allowable catch levels, and allocations means balancing a large number of interests that all make up the public interest. It requires "truly objective knowledgeable individuals who will bite the bullet" as council members. But the interests of some council members are narrow. "Council composition precludes free thinking. You get the constituent views instead of a broader approach to the issues," observes an agency administrator.

Another part of the problem is the potential for council members to have conflicts of interest. It would be possible, one manager thinks, to "reduce the special interest problem if council members were active in fisheries about which decisions are being made . . . but able to broaden their interest as time on the council passes. We need . . . quality people who have broad perspective . . . statesmanship ability." Some believe that the problem would be improved by holding the appointment process to a much higher standard of scrutiny and having stronger oversight by the secretary of commerce. Others recommend eliminating the common practice of campaigning for council seats. Campaigning can lead to the appointment of people with more political connections than knowledge and who are beholden to the interests most influential in their selection. But conflict of interest is not limited to public members, some say. State agency directors can have similarly strong conflicts of interest as they try to please their state legislatures and constituents.

Improved council accountability results from recording the votes of members. Some types of votes taken by councils are required to be roll-call votes, while others may be either voice votes or roll-call votes at the discretion of the council chair. Recording all votes in the same way that congressional votes are ranked is recommended by a council participant as a way to "hold their feet to the fire."

One person we interviewed suggests that council member compensation be tied to performance, with a system of bonuses tied to economic productivity or the health of the stock, depending on the management objectives. At the least, many argue, there should be stronger sanctions against councils for not meeting their management goals and objectives. But at the same time, many feel that councils are often not allowed to be responsible and accountable for their decisions because members of Congress are too anxious to intercede on behalf of constituents. They plead that Congress stop "making end runs" on the council system, because in doing so it is undermining the very system it put in place and eroding its ability to function as the regional system Congress intended it to be.

Others argue that all participants, not just council members, need to be more accountable. Strong statements are made at council meetings, and they are not always backed by evidence. The temptation to exaggerate is strongest in times of greatest conflict, but this is also the time when accurate information is at a premium to decisonmakers. Several people suggest that industry groups, environmental organizations, scientists, and all other participants be held to standards of truth and accountability for the information they present, perhaps by testifying under oath.

National Marine Fisheries Service personnel too are asked to be more accountable to constituents, to be given incentives to stay in touch with constituents and develop stronger ties with industry. One industry member expresses the frustration of many about trying to work with agency scientists. "They say, 'Why are you wasting time? Those guys don't listen.' " On the other hand, agencies are also criticized in our interviews for being too responsive to constituents and subverting the intent of regulations. There are clear differences among stakeholders about whether council members and agencies are too responsive or not responsive enough.

People also apply the need for accountability to the fishing industry and the positions it takes. To some, it is a matter of the industry having too much authority in the council system; to others, it is too little. Several observe that many industry members are disenchanted with councils. One environmental representative notes, "People have to be motivated, to be able to see change come out of their participation. Too often the industry advisory recommendations are completely ignored without any explanation, any obligation for the council to explain why." But, cautions another, "Making fishermen into bureaucrats won't work. We have to find a way for them to influence decisionmaking but remain fishermen."

A frustrated industry member points to the lack of knowledge among council members about the economics of fisheries and of fishery management tools. "Managers don't understand the economic dynamics going on here, so they work at cross purposes. For example, they make licenses nontradable to make them more politically palatable, but all that does is remove equity value from long-term participants in the fishery. If a man dies, leaving his boat and his business, his wife can't continue the business or sell the boat without a license, which died with the fisherman. The result is that because gutless fishery managers are unwilling to stand up and make hard decisions to reduce the number of licenses, widows and orphans pay to get the [number of] licenses down."

Would involving a broader community of people make decisionmaking more responsible? One academic stakeholder points to the potential benefits of involving people from outside the traditional fishery management structure to share solutions to similar problems in other fields. "There is a low potential for getting people who have a long involvement to change, but getting the public and people from other arenas involved can stimulate change," he says. "Innovation will come from outside the system." Another counters, "Management should be a little more dogmatic in telling people what to do, not asking them what they want to do. Management needs to step up to the plate and say we must make changes."

Some see effective management being enhanced by a careful balancing of the skills and knowledge of commercial fishermen, recreational fishermen, environmentalists, academics, and others. They criticize some councils for not using advisory panels well and for failing to take advantage of the many sources of knowledge among participants. Others focus on the potential benefits from more cooperative research and data collection between scientists and nonscientists. Many point out that fishermen, in particular, are natural innovators with many ideas that are not brought into management. But the degree to which industry knowledge is used clearly varies across councils, and our interviews acknowledge that some councils do effectively use advisory panels and are more effective than others in this regard.

An increasingly common recommendation for broadening participation has been for community-based management, although the understanding of how this would work for the full range of fisheries is still limited. Some see a community as a more appropriate scale at which to conduct research, design management, and enforce regulations. The regional councils "don't get down to a local enough scale," as one manager sees the problem. "The links are lost." A researcher supports this view, explaining, "When you're looking at things from such a large scale, things that occur at a smaller scale disappear from your view."

But not all agree. Says one representative of an environmental organization, "There is no sense to return management to communities adjacent to the fishery. It will only fragment management and cause problems. . . . Community-based management is a fun idea, and local communities love it. But it makes lousy science and lousy management. It doesn't make sense from the biological resource perspective because most resources are at a larger scale than a community. So if you manage over the full range of the ecosystem you are going beyond the bounds of the community. The scientific information is too complex for communities to manage. We would be giving away the resources of the American public." An industry member adds his concern about the use of community development quotas: "It locks in relations between communities, ports, and processors. . . . It's the ultimate command and control solution that leads to company towns. The sharecropper approach. . . . Community-based quotas will freeze the system in place. Why are we afraid of change or progress? It's bizarre. It's market power exercised through another channel. It's the status quo seeking to preserve itself."

Decisions can also be improved by research. "We can't change the status quo without improving research," says one environmentalist, who remarked on the need for research to better understand how to define and implement concepts like the precautionary approach. Research can reduce uncertainty about the condition of stocks, without which the precautionary approach will dictate more conservative total allowable catches. Research can also increase understanding of ecosystem interactions and how fishing affects them. An industry member argues, "We need a better overall understanding of the problems and how they were created—species interactions; pinniped [seal and sea lion] impacts; whether overharvest of pinniped forage fish has led them to change their predation patterns. One thing impacts another and we have very little information on those interactions."

Take the Long View

The development of trust in a long-term common future can reduce the incentives for getting as much fish as possible now.

Angela Sanfilippo, Gloucester Fishermen's Wives Association

We need urgently to develop programs that facilitate exit from the fisheries. We should allow people to exit gracefully, and get their kids equipped to have alternative jobs.

Jim Crutchfield, University of Washington emeritus

For the many reasons already discussed, fishery management is usually short term in its approach. As management has become more complicated and uncertain, the perspective has become even shorter. Councils are in a position of trying to meet regulatory requirements, contain fishing effort, and address conflicts and allocation requests. They have little time to step back and plan for the future, nor are they required to. Short-term requirements "distract people from what they should be doing," says a fishery manager. "We need to choose an issue and make it a focus point." "Councils are not the place to come up with new ideas," observes a researcher, perhaps in acknowledgment of the huge workload associated with routine annual work requirements.

Difficulty in taking the long view is common to almost everyone, and certainly affects fishery participants. If a fishery management plan extends too far out into the future, those in fishery management today will not see it come to fruition, and so it becomes meaningless to them. As a recreational representative observes, "The best way to achieve recreational buy-in to stock rebuilding plans is to rebuild within a reasonable period of time and to leave a substitute for the fishermen in the meantime. Recreational fishermen are more understanding when they see more of a balance between those who cause the problem and those who have to bear the burden of recovery and when the time horizon allows them to realize the benefits of cutting back."

To take a long-term perspective requires that people feel secure that they will be involved in the fishery for the long term. This applies to all involved in fishery management—to industry, anglers, environmentalists, managers, and scientists alike. If tenure in the fishery is secure, stability and predictability are enhanced. But securing tenure goes hand in hand with limiting participation to a defined group and denying tenure for others, and that is the source of the political difficulty in focusing on the long term. It is natural in times of economic difficulty to want to find ways to preserve the past by assisting the industry "until things turn around." But several we interviewed suggest that to try to artificially maintain what is no longer viable is doing the fishery few favors. Taking the long view, they say, we've done very little to address the incentives that cause overfishing and economic troubles.

Many think that, first and foremost, the fishery subsidies should be removed so that the costs of fishing are not kept artificially low. Eliminating these programs is not a simple matter either politically or financially. For example, as of 1995, the accounts of the Capital Construction Fund contained $242 million in tax-deferred income and it is infeasible to abruptly subject those monies to income taxes. Different ideas exist about what to do with these accounts if the program were ended. Several involve proposals to roll over the funds to other tax-deferred uses that would be more compatible with long-term goals for fisheries. One stakeholder proposes that we "convert the incentive to expand capital to expand research and management . . . [and] allow the money in the fund to be put into fisheries management or fisheries research." Other proposals include allowing money in a Capital Construction Fund account to be used for the purchase of individual fishing quotas.[13]

Also critical to the long term is the need to reduce capacity and resolve the problem of access so that future capacity can be maintained at levels the resource can support. Overcapacity is seen as both the cause and the effect of the inability of industry and managers to take a long-term perspective to management. While the recent trend has been to close fishery access, and open access is now a thing of the past for almost all federally managed fisheries, overcapacity and its associated pressures remain, keeping the focus on short-term attempts to contain overcapacity's effects on the fishery. Some government programs have bought back vessels or permits and more are proposed. Reducing the amount of fishing capacity on the water and in the management system is a precondition of being able to plan for the long term. "If that problem is not solved, then all bets are off," says one stakeholder, reflecting a common sentiment. At the same time, reducing capacity will not in itself be enough to end short-term thinking.

To many, the path to resolving the problem of overcapacity for the long term is rights-based management. This is the only management approach in which the interests of most players in the game come together. "A market system could let people figure out how they want to harvest the fish," says one stakeholder. Others in agreement with these sentiments argue for lifting the moratorium on the development of new individual fishing quota programs, saying that prohibiting councils from using them is a disincentive to dealing with the long-term problems of bringing capacity into balance with the resource.

Several note the importance of having councils plan ahead so that when capacity is reduced it is done with dignity rather than in crisis. It is already too late for some fisheries, where bankruptcies, vessel sinkings, and involuntary exit have been the methods of reduction.

Participate Responsibly

> What happens to fish happens on the boat, so it's people on the boat that really matter to the fish; how that gear is fished and interacts with them.
>
> *Robin Alden, formerly Maine Department of Marine Resources*

> The incentive for stewardship comes from giving fishermen a sense of ownership and decoupling the whole process of unbridled competition that is characteristic of open access fisheries. . . . A sense of ownership and responsibility is essential for stewardship.
>
> *Michael J. Fogarty, University of Maryland*

In U.S. fisheries, responsible participation has two facets—responsible fishing and responsible engagement in management. While all would agree that people should fish responsibly and that management based on participation requires responsible behavior, there are different views about what this means.[14]

Some focus on enhancing the quality of participation in management. They point to the need for better understanding among some council members of the scientific information they use to make decisions. Some argue for stable rules—noting that the rules change too often and are applied unfairly. "Industry participants need consistency in management philosophy and approach," says one. Others support this by pointing out how frequent changes in rules remove business security.

Many point out that the way fish are allocated by some councils affects the quality of participation. They note especially that when decisions are unguided by formal analysis of social and economic properties of the fishery, participants are encouraged to make exaggerated claims at council meetings to make the case for social and economic impacts. Still others blame the lack of responsible participation on enforcement systems that are often developed without input from fishermen.

In contrast, some note the positive participation that is already in place. From some industry voices comes the plea for greater recognition from managers for the contributions industry makes to problem solving. "No one will ever write the originators of those innovative ideas to say thank you," one says. "The system tells you all along what you can't do, not what you can. It doesn't reward service or ideas, particularly those of fishermen. A lot of ideas for improvements can come from fishermen."

To others, responsible fishing and responsible participation in management go hand in hand and require three things: assurance, rewards, and education. Fishery participants need assurance that a future exists— that the benefits of actions taken today will accrue to those who take them. "If we have to reduce landings of fish, we will gladly do it if we are also the ones who benefit from this cutback," says one fisherman by way of illustration. "This is a different impression of fishermen than many people have." A fishery observer agrees: "If they know they will survive economically, it is logical to take stewardship responsibilities seriously. When there is uncertainty, chaos, and anxiety regarding bankruptcy, there is no time for stewardship."

The question of assurance about the future is essential. Since under open access, as well as some limited access programs, fishermen have no assurance about their future stake in the fishery, they risk losing the benefits of today's conservation actions to others. This risk removes their incentive to conserve. "People have to be motivated to be able to see some change that comes out of their participation," one says. Until fishermen have a stronger stake in the future of the fishery, the incentives of fishing will be difficult to align with responsible fishing.

Stakeholders from all sectors point to the increase in assurance that would result from clarifying property rights. "Fishermen behave rationally given a management plan, but they are given incentives to go out and catch as many fish as they can. Until they are given incentives for conservation and rewarded for good behavior, we'll continue to see management failures," one environmentalist observes. Share-based systems directly reward people for healthier stocks because the size of their share depends on the size of the total allowable catch. The assurance of property rights lies in giving fishermen an incentive to think about the long-term benefits and find more effective ways to meet management objectives. "Innovations in operations and management won't come until we have rights-based management with vested interests," one academic says.

Related to assurance is the need to build more trust among industry members. Assurance also has to do with confidence that the rules will apply fairly to all. A researcher notes that "enforcement is currently based on strict fines, but it doesn't deal with the long-term status of violators in fisheries." To some, the profit motive embedded in rights-based management will provide the incentive for problem solving, including improving enforcement. "Fishing rights—that's all you need. I base that on my experience with the Canadian fisheries compared to the U.S. West Coast. In Canada, everything changed in two years. The people aren't different, but the incentives are all wrong without fishing rights."

The lack of any type of ownership is, to many, the fundamental problem for responsible participation in either fishing or management. "Without rights, you have nothing," argue many who propose rights-based management tools as the best way to reduce capacity, strengthen the stake in the fishery, and encourage stewardship behavior. But others believe that giving away rights to fish would do nothing to induce responsible fishing, citing concerns about short-term profit perspectives overriding long-term stewardship needs.

Many stakeholders see the problem of overcapacity as one that undermines all attempts to improve responsible participation because it provides incentives to race for fish or to fish "dirty" to maintain competitive advantage. Reducing overcapacity is the key to strengthening the assurance fishermen have about their future, particularly reducing capacity through incentives that allow people to "exit the fishery with dignity."

Rewards and penalties can influence responsible participation as much as assurance can. The focus of regulation has been on penalties for breaking the rules. Rewarding good behavior can be just as important. Several industry members think that rewarding good fishing practices is the best way to clean up the fisheries. "If you work hard to save fish you should reap some of the rewards" is a common industry view. With individual accountability, fishermen could use their own ingenuity to solve a problem and could then be rewarded with additional days at sea or increased trip limits. Those boats with the lowest bycatch rates would be considered the "responsible fishers" who would qualify for extra benefits, although, if the benefits are expressed in terms of increased quantities of fish or fishing time, there is an obvious limit to using this approach.

Penalties for rules violations are slow and often unfair, several industry members lament. Not all rules are enforceable, and fishery managers are seen by some to be too soft on certain industry sectors and bad actors. The incentive to do more self-enforcement that derives from restricted access would help, say many. So would fairness in enforcement. "Fishermen decide not to cheat because they're pretty sure that the others aren't cheating also," says one scientist. That kind of assurance is difficult to find under the current structure, he argues, saying that it's only developed in small, well-integrated groups. "People have to know that they're not the only idiot out there who is following the rules. It's easy to cheat when you're out there on your own and there's nobody around."

To others, responsible behavior is a matter of education and a condition of qualification. "If entry is slowed, based on qualifications, for example, an apprenticeship process that teaches or provides tools for stewardship and requires an investment of human beings' time, not just money, you're giving a chance for stewardship," says one stakeholder, referring to Maine's experiment with apprenticeship programs in the lobster fishery.

Perhaps the most important point about improving the participation of fishermen has to do with the need to extend self-interest from the short to the long term. An industry member observes that the industry's orientation is very short term. "It shapes the view of 'I prefer possible bankruptcy tomorrow to certain bankruptcy today.' It's not a wrong view—it's what it is. In contrast, some environmentalists have excessively long time horizons. They are looking forward 30 years from now and tend to forget what happens to people in the meantime, in the short term."

A researcher, noting that the power to exclude provides incentives to conserve, extended that realization to a requirement to commit to a particular fishery. "Fisheries ought to be regional, with the same people going back year after year, versus a highly mobile fleet, opportunistically zipping here and there, fishing places out and then moving on. The system should require people to make a commitment to one fishery, so if they screw it up by using bad fishing practices, that's it. They'll have to live with the consequences of their actions."

"But fishermen are changing," says one industry member, "they are more educated and most have lost the mentality of filling the boat, no matter what." What is often needed is education. An environmentalist also notes this need as well as the need for financial assistance: "We demand that fishers comply with gear restrictions but we never establish programs to assist them." Another extends the need for outreach education for the large public. "Too many people, although living close to the coast, have no connection with the ecosystem," he says.

Finally, many stakeholders focus on ways to make better use of the fishery's human capital, saying that more involved participation could also improve decisionmaking. Industry members across regions are requesting a greater involvement in fishery research in order to contribute to better quality data and expand their fishing opportunities. Many cite the management benefits of conducting more joint research with both the commercial and the recreational industries. These benefits would be educational, as industry members and scientists learn more about each other's methods and would also contribute, in the longer term, to the integrity and legitimacy of data used in management. "We need greater involvement by the fishing industry in scientific research so the fish become more than just a commodity to prey upon," asserts one scientist. "We need greater outreach by fishery scientists to train the fishing industry in the methods of science and the limitations of casual observation."

A former government administrator argues that the way to get constituents, and particularly industry, to participate more effectively is to encourage their cooperation in the collection of data and to involve them more directly in "planning, conducting, and evaluating projects and developing recommendations. This has to be more than just lip service as it frequently is now. People will want to be on the team if they are truly listened to, the information is evaluated and used as justified, and the results make a difference." Others see an equal need for scientists to learn from the ecological knowledge gained by the fishing industry through constant interaction with the ocean environment.

Industry members and fishery scientists do collaborate on a number of ongoing research projects. These programs integrate industry operations with scientific research through joint support and conduct of projects. State fishery agencies and the National Marine Fisheries Service are involved in a number of collaborative efforts with industry and more are planned. In addition, the Saltonstall-Kennedy grants program for industry-related research directly involves industry as collaborators and principal investigators. A multitude of programs has assisted fishermen in complying with new gear regulations. The Sea Grant Marine Advisory Program also works closely with industry to generate new knowledge.

Conclusion

Right now, those who stand to gain the most have put the burden of proof on the managers and scientists. I would turn that around and have the managers and scientists put the burden of proof on those who stand to gain the most.

Paul Howard, New England Fishery Management Council

When you remove the fishers' responsibility, you remove incentives to create and follow appropriate rules.

James Wilson, University of Maine

One of the most hopeful signs for fishery management is that people are, for the most part, rational in what they do. They respond to the incentives that face them and, to the extent that these incentives are understood, their behavior is predictable. Unacceptable outcomes in fishery management occur not necessarily because people are greedy or stupid, but often because of the incentives created by fishery policies and decision processes. The incentives created by open access and the race for fish have encouraged managers to take the short-term view and fishermen to act against their own long-term interest by investing in too much capacity and fishing "dirty" with too much bycatch.

If we are not achieving the goals of fishery management, what is standing in the way? How can we turn the incentives around to promote the behavior we want? "The basic algorithm is throw back the little ones, leave some for tomorrow—what you were told when you were a kid," says an environmentalist. How do we achieve this so that healthy fisheries are maintained, responsible decisions are made, the long view is taken, and responsible participation is promoted?

Incentives need to be brought into line with the current goals for U.S. fisheries. Although people have different opinions about the specific forms the changes should take, most agree that the fundamental needs are to remove the incentive to increase fishing capacity, provide incentives to reduce capacity, clarify rights of access, and develop systems of rewards and accountability for responsible decisionmaking, collaborative research, and stewardship.

In Their Own Words*

*These statements reflect the personal views of interview participants and do not necessarily represent the positions of the institutions or organizations with which they are associated.

Scott Burns, Director, Marine Conservation Program, World Wildlife Fund

We've done very little to try to address the incentives that drive overfishing. We need to develop effective plans to deal with overcapacity. This should include a thorough look at how to limit access, including rights-based management. We should also look at purchasers to try to get them to be more selective in what they buy. That would send a very effective signal. We need to broaden our horizons though and recognize that our fisheries are competing in a global market. If tough restrictions are imposed on fishermen here, the same should be done elsewhere so that our fishermen are able to compete. We can't focus on the United States and ignore what's going on elsewhere—this could put our fishermen out of business.

Christopher M. Dewees, Marine Fisheries Specialist and Sea Grant Extension Program Coordinator, University of California, Davis

We've lost the opportunity to examine alternative ways of managing fisheries. Barriers to considering alternatives are so great that it's hard to take seriously or try new things—like cooperative or zonal management or protected area-based management strategies. We've lost the ability to do things in a flexible way. Why? Fear of change, concern about being a loser, lack of political will by government, and, possibly, even some aspects of the Magnuson-Stevens Act could contribute.

Vince O'Shea, Captain, U.S. Coast Guard, 17th District, Juneau, Alaska

Realistically, enforcement cannot correct "mass behavior." Fundamentally, management measures must be supported by resource users. One of the important strengths of the current fishery management council system is that it contains the tools of public process to help ensure buy-in by the participants. Having the majority of resource users agreeing with the management plan means that enforcement resources can concentrate on the fringe element that tries to operate outside the system. In this context, enforcement provides several important services to the public. It helps ensure that fisheries resources are protected, it provides confidence in the operating integrity of the management system, and it provides a level playing field by preventing cheaters from enjoying windfall gains.

Angela Sanfilippo, President, Gloucester Fishermen's Wives Association

There are currently no positive incentives to reward responsible fishing practices, such as additional days at sea, increased trip limits, higher dockside prices, or less red tape.

There are only penalties, fines, and confiscations to punish offenders and these are not even applied equally. People who break the law get away with it for years and then they are rewarded with greater quota or more days at sea. Pleas from the industry to make rules that are enforceable are ignored, rules fail to meet their goals, and the vicious cycle continues its downward spin. Eventually the dishonest "fishcats" own it all. We need to make sure that rules are enforceable and are enforced equally. All evidence indicates that a grass-roots incentive approach would be more successful than the present top-down system that rewards the wild west approach and the dishonest crooks.

James Wilson, Associate Director, School of Marine Sciences, and Professor of Resource Economics, University of Maine

We have to find a way to create an atmosphere in which fishermen find it in their interest to avoid things like throwing a net into a nursery or spawning area. Because, until they see themselves as stewards, we'll have to put a policeman aboard every boat. Incentives for stewardship will result from increasing fishermen's responsibility, autonomy, and independence. And, should fishermen develop a sense of stewardship, the question is then, How do we bring other interests into the process without destroying it? If environmental and recreational groups override their decisions we'll go right back to a top-down approach and those groups will lose what they want because fishers will end up degrading the system. We have to find a reasonable and equitable way to let other interests have a say in the resource.

Notes

1. Milazzo, M. 1998. *Subsidies in World Fisheries: A Reexamination.* World Bank Technical Paper, Fisheries Series, No. 406. Washington, DC.
2. Pauly, D., V. Christensen, J. Dalsgaard, R. Froese, and F. Torres Jr. 1998. Fishing Down Marine Food Webs. *Science* 279:860–63.
3. Hanna, S. 1997. The New Frontier of American Fisheries Governance. *Ecological Economics* 20:221–33.
4. Hanna, S. 1998. Institutions for Marine Ecosystems: Economic Incentives and Fishery Management. *Ecological Applications* 8(1) Supplement:S170–S174.

5. NMFS. 1996. *Magnuson-Stevens Fishery Conservation and Management Act.* NOAA Technical Memorandum NMFS-F/SPO-23, December.

6. Roodman, D. M. 1996. *Paying the Piper: Subsidies, Politics and the Environment.* Worldwatch Paper 133. Worldwatch Institute, Washington, DC.

7. Myers, N. 1998. *Perverse Subsidies.* International Institute for Sustainable Development, Winnipeg, Manitoba, Canada.

8. Iudicello, S., M. Weber, and R. Wieland. 1999. *Fish, Markets and Fishermen: The Economics of Overfishing.* Island Press, Washington, DC.

9. NMFS. 1996. *Our Living Oceans. The Economic Status of U.S. Fisheries.* NOAA Technical Memorandum NMFS-F/SPO-22, December.

10. Ibid.

11. Federal Fisheries Investment Task Force. 1999. Report to Congress. July.

12. Ibid.

13. Ibid.

14. Wilson, D., and B. J. McCay. 1998. How The Participants Talk About "Participation" in Mid-Atlantic Fisheries Management. *Ocean and Coastal Management* 41:41–69.

CHAPTER 7

Using Scientific Information

The most profound weakness is the politicization of the fishery which causes the science to be ignored. Management becomes a fight over allocation rather than conservation.

Cynthia Sarthou, Gulf Restoration Network

Politics trump science every time.

William T. Burke, University of Washington

The MSFCMA codifies the principles of fishery conservation and management in the form of 10 national standards (listed in chapter 1). The second national standard dictates that fishery "conservation and management measures shall be based upon the best scientific information available."[1] Yet relying on the best scientific information available and being cautious in its use does not mean that fishery management measures will always be fully informed. Marine ecosystems are complex and dynamic, as are the people who use them. As a consequence, large knowledge gaps and considerable uncertainty surround the decisions made in fishery management.

Fishery management decisions rely—at least in theory—on scientific information that encompasses all biological and human components of the fishery system. Biological information, primarily derived from at-sea surveys and samples of landed catch, is used to assess the status of fish stocks. Economic and social information, based on recreational surveys, sales of landed catch, vessel registration files, and surveys of vessels, plants, and people, describes the human components of the fishery, including those who catch, process, and sell the fish, and the communities within which they live.

Does the best scientific information available to fishery managers adequately describe these components of the fisheries? Is the best available information enough? It is always easy to say that more information is needed. But what do we need the most? Is existing information generated in a cost-effective way, used wisely, and communicated effectively? Does it address the questions being raised? If not, what might be done?

Why Do We Need Science?

There's never enough science to eliminate the politics.

John Green, formerly Texas Parks and Wildlife Department

The science is very often good enough to make reasonably well-informed decisions. It is never good enough to be invulnerable to attack.

Ken Hinman, National Coalition for Marine Conservation

The first national standard—and the heart of fishery management—states that "conservation and management measures shall prevent overfishing while achieving, on a continuing basis, the optimum yield from each fishery for the United States fishing industry."[2] Fishery management plans must "contain the conservation and management measures . . . which are necessary and appropriate for the conservation and management of the fishery to prevent overfishing and rebuild overfished stocks, and to protect, restore, and promote the long-term health and stability of the fishery."[3]

To achieve this directive, fishery managers are required to define objective and measurable criteria that will indicate when a stock is being overfished. Such criteria could be based on measures of fishing mortality or spawning potential and are determined through biological stock assessments. But stock assessments provide only part of the information that is needed for biological fishery management. The act, as of 1996, also requires managers to describe and identify the habitat necessary to sustain each stock, to minimize the effects of fishing on such habitat, and to include measures to minimize bycatch, bycatch mortality, and catch-and-release mortality to the extent practicable.

These requirements focus on the conservation and management of fishery resources. But what about the people who make up the fishery? "Only half or fewer of council decisions relate to biological issues," one scientific adviser notes. "The rest has to do with allocation, which requires social and economic data." The MSFCMA recognizes this by providing fishery managers with multiple directives to assess the economic, social, and cultural status of fisheries as well.

Specifically, the act requires that each fishery management plan describe social and economic aspects of the commercial, recreational, and charter fishing sectors that participate in the fishery, in terms of the number of vessels involved, the type and quantity of fishing gear used, the costs likely to be incurred in management, and actual and potential revenues from the fishery. It also calls for a description of any applicable Indian treaty fishing rights and states that restrictions and recovery benefits from catch reductions must be distributed equitably among competing interests. Fishery management plans and amendments dated after October 1, 1990, must include a fishery impact statement that assesses, specifies, and describes the likely effects of management measures on these sectors and fishing communities.

The eighth national standard requires that conservation and management measures—without sacrificing conservation goals—take into consideration the importance of fishery resources to fishing communities, provide for these communities' sustained participation, and minimize adverse economic impacts on them to the extent practicable. Examples of such impacts are provided in the National Marine Fisheries Service's national standard guidelines and include decreased employment opportunities in the fishing and processing sectors and inequitable allocation of fishery resources.[4]

Social and economic data and analyses are needed not only to meet these MSFCMA requirements, but to satisfy other legislation as well, such as the Small Business Act, the National Environmental Policy Act, the Regulatory Flexibility Act, and Executive Order 12866: Regulatory Planning and Review.

Enacted in 1953, the Small Business Act requires that agencies assist and protect small-business interests to the extent possible to preserve free competitive enterprise. The National Environmental Policy Act of 1969 and implementing regulations require agencies to consider the potential environmental and human consequences of policy and regulations through the development of environmental impact statements. Social considerations are to be accounted for through the development of social impact assessments but, although the National Marine Fisheries Service produced a document in 1994 to guide such assessments,[5] they do not appear to be a regular component of fishery management plans. And those that are prepared are often incomplete.[6]

The Regulatory Flexibility Act of 1980 instructs agencies to minimize adverse impacts from burdensome regulations and record-keeping requirements on small businesses, small organizations, and small governmental entities. It requires that proposed rules be examined as to whether they will have a significant economic impact on a substantial number of small entities. If not, a certification to this effect must be prepared. Alternatively, if a proposed rule will significantly impact a substantial number of small entities, the act requires the completion of a regulatory flexibility analysis for public comment. Such analyses describe the type and number of small businesses affected, the nature and size of the impacts, and alternatives that minimize these impacts while accomplishing the stated objectives. The analyses must be published in the Federal Register in full or in summary and submitted to the chief counsel for advocacy of the Small Business Administration. They become final after public comments have been addressed.

Finally, Executive Order 12866 signed in 1993 provides guidelines to ensure that agency regulations are efficient and cost-effective. It requires agencies to consider the costs and benefits of management measures, including factors such as equity and the distribution of impacts among different people. A regulation is considered significant if it is likely to result in an annual effect on the economy of at least $100 million or has other major economic effects.

In compliance with Executive Order 12866, the National Marine Fisheries Service requires the preparation of a regulatory impact review for all fishery regulatory actions that either implement a new fishery management plan or significantly amend an existing plan. These reviews provide a comprehensive analysis of the costs and benefits to society associated with proposed regulatory actions, the problems and policy objectives prompting the regulatory proposals, and the major alternatives that could be used to solve the problems. The purpose of the reviews is to ensure that all available alternatives are considered so that the public can benefit in the most efficient and cost-effective way. They also serve as the basis for determining whether proposed regulations will have a significant economic impact on a substantial number of small entities in compliance with the Regulatory Flexibility Act.

Successful compliance with the extensive information and procedural requirements described above demands not only biological and social baseline data, but continuous monitoring as well. The MSFCMA also directs the secretary of commerce, in cooperation with the councils, to initiate and maintain a comprehensive program of fishery research on these components of the fisheries. It also directs the secretary to publish in the Federal Register, at least every three years, a strategic plan that outlines the course of fisheries research five years in advance.

Is the Science Adequate?

Because fisheries include both fish stocks and people, the scientific basis of management decisions must include biological information on

fish populations and ecological communities, as well as human information on economies, culture, and social relationships. These are the
realms of biological and social science. How adequate is the biological
and social scientific information base for management?

Biological Science

> The councils are operating with inadequate data. Stock assess-
> ments need to be improved.
>
> *Roger Thomas, Golden Gate Fishermen's Association*

> The heyday of fishery research was when researchers went out on
> fishing boats and measured fish and kept track of catch per tow. . . .
> It is my view that the management process could work better if
> actual fisheries data were used more often.
>
> *Brian J. Rothschild, University of Massachusetts*

Most fishery stakeholders share the view that, overall, the biological
stock assessments that inform fishery management decisions are imperfect, yet adequate to make reasonable management decisions. But this
view varies among regions, fisheries, and sectors.

Many argue that stock assessments could greatly benefit from improvements in the capacity for real-time data collection and analysis and
the incorporation of recreational fishing data. Some believe that more
local-scale research is needed so that conclusions from one area are not
extended to another, as is done for some species in the Gulf of Maine
from assessments done on Georges Bank, and in other areas as well. Others are concerned that, although capable of indicating with reasonable
accuracy whether stocks are increasing or decreasing, stock assessments
are incapable of adequately relating stock size and population dynamics
to benchmarks such as maximum sustainable yield or virgin stock size,
as required by the MSFCMA.

Almost all of those interviewed share a common concern that the status of two-thirds of the nation's fish stocks—the majority of which are
contained in fishery management plans—is still unknown.[7] According
to one, this is "unconscionable." But another points out that these stocks
for which status is unknown represent less than 3 percent of total U.S.
landings.

A few, particularly those whose futures are directly affected by decisions based on such assessments, strongly question the credibility of
stock assessments altogether. This perspective is best voiced by a New
England industry representative, who complains, "[Scientists] can't
even accurately count whales, but they are convinced that they can estimate the abundance of every fish species and a safe level of harvesting."

Supporters of stock assessment science counter that scientists are well-aware of the limitations of stock assessment data and models but—in spite of these limitations—are required to come up with best estimates of stock size. One argues that scientists "are routinely forced to extrapolate beyond the levels of precision for which the science is designed." All things considered, supporters point out, stock assessment scientists do a great job of addressing a complex problem. At the least, one argues, they have accurately characterized overfishing and stock declines in most cases and have done an admirable job of characterizing uncertainty and projecting the consequences of risk-prone management strategies.

Strong opposition by some to the use of stock assessment data has led Congress—on several occasions—to commission studies of biological fish stock assessments. Two recent studies by the National Research Council[8,9] found that, despite great controversy about the adequacy of these assessments, the process "appears to provide a valid scientific context for evaluating the status of fish populations and the effects of fishery management."[10] These reports also provide recommendations for improvements, such as evaluating research design, sample size, and data collection; encouraging experimentation with alternative methods and models; and including uncertainty in forecasting.[11] Many of those we interviewed support the latter recommendation, claiming that stock assessment science does not adequately account for environmental variability. But one scientist counters that techniques for incorporating environmental variability have a long history and are currently commonplace in fisheries science.

Many fishery stakeholders note a critical gap in information needed for ecosystem management, including information on basic life history traits of managed species. Even when life history information on targeted species is strong, they say, information on ecological relationships and essential habitat requirements is weak. New legislative and management requirements are increasing the need for such information, as well as a better understanding of the effects of fishing on the environment.

Congress recognized the importance, and challenge, of moving toward an ecosystem-based management approach in 1996, through an MSFCMA provision requesting that the secretary of commerce convene an advisory panel to make recommendations on the application of "ecosystem principles" to fishery management. One recommendation of the Ecosystem Principles Advisory Panel subsequently formed by the National Marine Fisheries Service suggests the establishment of marine protected areas.[12]

Concern for ecosystem protection, along with a desire for baseline ecological information, has helped the use of marine protected areas as a fishery management tool gain wide support among environmentalists and many academics. A few have even suggested formal legislation of marine protected area networks. Supporters view the increased use of this tool as a natural consequence of the move toward habitat and area management. They indicate that, acting as scientific controls, marine protected areas can provide valuable information on the health of marine ecosystems, while conserving the entire marine ecosystem and providing a buffer against future problems.

But others remain unconvinced about the conservation benefits of marine protected areas. Skeptics are concerned that frustration with the use of science in the fishery management process has caused fishery stakeholders to turn to widespread use of marine protected areas as an alternative to comprehensive, scientific planning. Although these areas are a promising new fishery management tool, they say, the benefits of marine protected areas remain untested. They note that marine protected area management can impose extremely high costs on resource users and that the design, use, and integration of such a tool with other fishery management measures must be carefully planned. In its 1999 report to Congress, the Ecosystem Principles Advisory Panel agreed that marine protected areas need to be further studied, noting that "research is needed on the optimal size of [marine protected areas], sources and sinks for new recruits, and the social and management issues required for successful implementation."[13]

Social Science

> When the "best available scientific information" was nothing, especially in economics and sociology, that's when I decided to exit the council process.
>
> *Robert B. Ditton, Texas A&M University*

> When making decisions that affect people's livelihoods, you would always like better data.
>
> *Lee G. Anderson, University of Delaware*

Many fishery stakeholders share the view that the social and economic aspects of fisheries and fishery management systems are not adequately assessed. In the words of one agency scientist, "economics and social science are impoverished." "Socioeconomic information is in the dark ages," another government official echoes. These views are based on a perceived disconnect between the social and economic information that the fishery management process requires and that which is produced.

Because little analytical information is produced about key aspects of fisheries related to fishery participants and their communities, many believe that the data necessary to assess these aspects of the fisheries simply do not exist. It is true that laws protecting business interests restrict the collection of socioeconomic data to some extent, making it difficult to obtain individual operator information. Even the MSFCMA bans the collection of proprietary economic information. But it is important to note that there are social and economic data routinely recorded through logbooks, fishing license applications, and other measures that are not routinely or systematically analyzed and summarized in a way that makes them useful, or applicable, to management decisions.

The National Marine Fisheries Service's Fisheries Statistics and Economics Division regularly produces statistical information on U.S. fisheries in the form of various bulletins, electronic information, and annual reports.[14] These documents include a great deal of information on U.S. commercial and marine recreational fishing, world fisheries, and the manufacture and commerce of fishery products. But they do not include information on social aspects of the fisheries, such as the nonmarket, nonuse, cultural, aesthetic, and existence values, which are difficult to measure. In addition, without routine, formal interpretation, or analysis, it is very difficult for fishery managers and scientists to apply the important information that is provided in these documents to management decisions.

Underutilized data are not a problem unique to fisheries. The Council on Environmental Quality recently concluded in a report on the effectiveness of the National Environmental Policy Act: "In addition to gathering more and better data, National Environmental Policy Act practitioners need to analyze existing information more effectively. . . . What is often lacking in environmental impact statements is not raw data, but meaning—i.e., a comparison of the potential impacts of choosing particular alternatives at particular locations expressed in clear, concise language. The National Environmental Policy Act is about making choices, not endlessly collecting raw data."[15]

How Is Scientific Information Used?

> Good scientists are becoming very frustrated with how science is being used, and misused, in the decisionmaking process.
>
> *J. Walter Milon, University of Florida*

> People do not want to hear bad news. When science produces that, science becomes the target of management.
>
> *Dick Schaefer, National Marine Fisheries Service*

The use of scientific information—both biological and social—varies greatly among the regional fishery management councils. Some use it quite extensively, allocating resources to data collection and analysis and relying on the scientific review and advice of their scientific and statistical committees. An observer of the North Pacific Council provides that council as one such example, indicating that its members have "accepted without argument the numbers the Scientific and Statistical Committee has given them." In extreme contrast, a fishery stakeholder in the northeast notes that the New England Council voted last year to reconvene its scientific and statistical committee for the first time since 1991. Others in the Mid-Atlantic and other council regions indicate that scientific advice is taken into consideration but can easily be rejected. One fishery manager notes, "We have always used, and put into fishery management plans, the best available scientific information. In some cases, however, that's only one document or study. So the best available scientific information is always debatable." It is still the case, another administrator claims, that legal advice carries more weight than scientific advice. In the words of another government official: "The role of science in fishery management seems less and less relevant as politics becomes more of the driving factor."

For this reason, some fishery stakeholders respond that the question of scientific adequacy may be the wrong question altogether. "This is a recipe for paralysis because the science will never be adequate," an environmentalist says. "We need to recognize that science has limits."

Biological Science

> The inherent uncertainty in fishery science has created an opportunity for strategic behavior based on that uncertainty.
>
> Timothy Hennessey, University of Rhode Island

> Too little science is always used as a roadblock to changing the status quo. . . . We need to recognize that science has limits. We have to make conservation choices in the face of uncertainty.
>
> Dorothy Childers, Alaska Marine Conservation Council

Biological information about fish stocks has a high degree of variability, making it difficult for fishery managers to accept such information, or resulting recommendations, at face value. There are two ways to respond. Managers can either use this uncertainty as an excuse to avoid taking actions that are politically unpopular, or they can take a precautionary approach, buffering their decisions to err on the side of caution. Choosing inaction in an uncertain and changing world, particularly when the resilience of the fish stocks has been reduced by heavy fishing, is thought by some to be a major cause of fisheries problems to date.[16] Most stakeholders agree.

In times of management distress, particularly when extreme cutbacks in fishing are necessary, science can become the "whipping boy" for other sources of dissatisfaction, such as the economic impacts of restrictive regulations. As described by one scientist, some tend to use uncertainty "as part of an end run game—ask for yet another peer review if you don't like the outcome." A recent National Research Council report on northeastern stock assessments supports this characterization, noting that "stock assessment science is not the real source of contention in the management of New England groundfish fisheries."[17]

Questioning the accuracy of science makes it easier for managers to ignore scientific advice that runs contrary to political goals. This is an enormous problem in fishery management according to fishery stakeholders in almost every region. As an industry representative on the Pacific coast describes, "The availability of scientific data is simply a hurdle to be overcome by council members whose votes are determined before the analysis begins." A council participant in the Northeast describes the common response to unwanted scientific information: "The science is out of whack, I'll just call my Congressman."

Even in cases where the science is fully adequate to inform a management decision, the political system can engender distrust. When science indicates that fish stocks are in decline, many welcome uncertainty as a rationale for denying the accuracy of the science and avoiding strict regulations to halt the decline.[18] When in doubt, this uncertainty makes it easier for some to accept estimates with the least impact on the fishery or split the difference. To this argument, an environmentalist counters, "Too little science is always used as a roadblock to changing the status quo. . . . We need to recognize that science has limits. We have to make conservation choices in the face of uncertainty . . . what level of science justifies the fishery? We can't wait for perfect science to justify change."

This is the basis of the precautionary approach, an approach strongly advocated by the conservation community, as well as by many scientists and fishery managers. The need for such an approach has been evidenced by humbling "surprises," such as the collapse of the northern cod stocks off Newfoundland and Labrador, Canada.[19] It is based on the principle that the limits of our knowledge require being cautious about levels of fishing.

The United States adopted the precautionary approach in both the U.N. Agreement on Highly Migratory Species and Straddling Stocks and the U.N. FAO Code of Conduct for Responsible Fisheries.[20] The MSFCMA also recognizes this approach in its definition of optimum yield and its requirements that councils define overfishing and restore overfished stocks to levels defined in precautionary terms. The recent National Research Council review of New England stock assessments for cod, haddock, and yellowtail flounder advocates this approach as well, concluding that when stocks are at low levels, high levels of fishing mortality increase the risk of irrevocable damage, requiring a precautionary approach.[21] Although guidelines have been developed to assist fishery managers in the use of such an approach in implementing the first national standard,[22] councils are not legally required to take a precautionary approach in their determination of allowable catch levels.

Social Science

> [National Marine Fisheries Service] data and data collection are woefully inadequate. We have a few random studies but we're having real trouble trying to even meet the requirements for social impact analyses. We're not able to do that for most fisheries or to determine community dependence on fisheries in the way that the act intends.
>
> *Peter Fricke, National Marine Fisheries Service*

The role of social science in council decisionmaking is also quite problematic, but for different reasons. In the absence of credible social science derived from a structured, scientific process, fishery managers are forced to assess social components at public hearings, where people bring biased perceptions. Testimony from the council floor is usually far more compelling than more formal social science analyses, one council participant notes. This is undoubtedly true. It is difficult for managers to ignore the heartfelt anguish of those negatively affected by regulations. But the danger of relying on anecdotes is that they present only piecemeal information that does not contribute to a systematic understanding of social ramifications. "The saying 'history is too important to be left for memory' is relevant," one respondent notes. The absence of a structured framework for systematically collecting, analyzing, and interpreting social science information also makes it difficult for fishery managers to fulfill their legal obligations. And, at times, they have been challenged by the courts.

In *North Carolina Fisheries Association v. Daley*,[23] the court held that the National Marine Fisheries Service had violated the Regulatory

Flexibility Act and the eighth national standard of the MSFCMA in setting its 1997 quota for the summer flounder fishery. It remanded the
quota issue to the Department of Commerce with an order to "undertake enough analysis to determine whether the quota had a significant
economic impact on the North Carolina fishery" and to "include in . . .
[the] analysis whether the adjusted quota will have a significant economic impact on a number of small entities in North Carolina."[24] Subsequent proceedings found that the economic analysis had failed to
give meaningful consideration to the economic impact of the decision
on North Carolina fishing communities.

One year later, the National Marine Fisheries Service procedures that
led to a 50 percent reduction in Atlantic shark catch quotas were challenged in *Southern Offshore Fishing Association v. Daley.*[25] In that case,
the Court concluded that the catch quotas were based on the best scientific information available but that the National Marine Fisheries Service's compliance with the Regulatory Flexibility Act was flawed. The
agency was instructed to undertake "a rational consideration of the
economic effects and potential alternatives to the 1997 quotas."[26] In so
doing, the National Marine Fisheries Service concluded that the catch
reduction may have significantly impacted a substantial number of
small entities, but that any alternative action would have jeopardized
the long-term viability of the fishery. The Court responded to the
agency's inability to develop practical alternatives with criticism, and
the case is presently before a mediator.

Social science information can best be incorporated in the fishery
management process by ensuring that data are collected, that plan development teams are interdisciplinary, and that social scientists are
involved in decisionmaking processes early on, when it is still possible
to inform the outcome. At present, some councils appear to be taking a
closer look at the use of social science information in their regions. Just
last year, the New England Council formed a social science advisory
committee to evaluate the socioeconomic data and analyses contained in
regional fishery management plans and to recommend improvements. A
participant in the North Pacific region notes that the North Pacific
Council has done the same and is now working to develop protocols for
obtaining data not previously collected. But, despite this increased
attention to the incorporation of socioeconomic information in the fishery management process, the system still lacks critical resources.
According to a former National Marine Fisheries Service administrator,
"We need social scientists and economists in every region and we don't
have them now. We also need them in Headquarters."

Communicating Scientific Information

> From my 13 years on the [scientific and statistical committee] I have learned to appreciate the idea of scientific advice being discussed in a public forum and the usefulness of having different sources of information; for example, observations from the fishing grounds that may explain model results, influence model assumptions, or predict the effectiveness of regulations.
>
> *Terrance J. Quinn II, University of Alaska*

> I'd like to see those who do good research, and fishery managers who do credible jobs, receive awards.
>
> *Barry Fisher, Midwater Trawlers Cooperative*

In a management system based on participation and science, the communication of scientific information is extremely important. As one environmentalist notes, "It doesn't matter if you're right if no one agrees with you." Scientists of all kinds need to translate their information into forms that others can understand and use in decisionmaking. Industry, in turn, needs to present information in ways that are relevant to the scientific process.

At present, the communication of scientific information is less effective than it could be in many regions. Some cannot even agree on the extent to which such information should be communicated. In the Northeast, one government scientist claims that the fisheries science center is a "pioneer" in the United States in terms of the public participation and peer review process that it has initiated for biological stock assessments. But another scientist in that region disagrees, saying that the "science is more sophisticated, but not always visible to constituents." An environmentalist refers to the National Marine Fisheries Service approach to science in the northeast region as "mandarin policy," essentially closed to outside review or outside sources of information, where models generate "numbers no one believes."

Such criticisms are not confined to the East Coast. Stakeholders on the West Coast also express great frustration with what they consider to be a closed process of data collection and analysis, referring to the resulting information as a "little bit of survey data tweaked by modelers." This feeling persists despite the Pacific Fishery Management Council's efforts to make scientific information widely available at council meetings, in newsletters, on the council Web site, and through open scientific reviews of stock assessments. The North Pacific is a region where scientists have done a better job of communicating what the science issues are and how science should be treated, according to an academic researcher. In his view, this communication has led nonscientists to a greater understanding of the issues and to the improved use of science in fishery management decisions.

Fishery stakeholders provide different explanations for the generally poor communication of scientific information. One academic observes that "scientific information is typically not well-formatted for use by decisionmakers," while a government scientist suggests that some of the confusion is strategic, deliberate, and related to allocation. Another academic faults fisheries scientists for their poor ability to incorporate uncertainty in scientific advice and present this uncertainty to fishery managers and policymakers. Yet another counters that biological scientists have made huge strides in modeling and presenting uncertainty over the last two decades, but that fishery managers continually use this uncertainty to preserve status quo catch levels.

Others believe that communication is affected by the antagonistic relationships that develop among participants, which works against the open exchange of information. For example, the "siege mentality" that develops on the part of some scientists in the face of criticism is often expressed in attitudes and behaviors that are perceived by others as arrogance. This exacerbates already large differences in values among scientists, managers, and industry members. A fisherman, reflecting on his experience with a council, says, "We hear something ridiculous and think they [the scientists] are stupid, and they don't give us [the industry] due respect."

One government scientist argues that, to improve communication, it is important that "people who understand science are at the table [during discussions and management planning] to prevent misinterpretation and misunderstanding." Those with long institutional memories and more experience with the issues provide an added benefit, the researcher adds. Many others argue that science should be completely separated from the process by which decisions are made. One approach, suggested by an academic, would be modeled on the Environmental Protection Agency's scientific advisory boards. These boards operate independently of the regulatory process, with more autonomy than the scientific and statistical committees. Another academic proposes the creation of a separate corps of professionals trained in new methods of translating scientific information into a format that is useful for decisionmakers.

Funding Science

> We seem to have adopted a national policy of providing inadequate
> funding for federal fisheries management.
>
> *Burr Heneman, marine policy consultant*

> The ocean is an expensive place to do research and Congress has
> not been willing to put up the money to do it. This is a big need.
>
> *William J. Chandler, National Parks Conservation Association*

Whether arguing for better biological, ecological, economic, or social
scientific information, the most frequent recommendation among fishery stakeholders from all sectors is for increased funding. This means
increases from the various sources of funding that contribute to fishery
management. Congressional appropriations fund the National Marine
Fisheries Service to conduct in-house and contract research. States provide funds through their fishery agencies. State and federal funds support Sea Grant College Programs, which include both academic research
and outreach education of research results. User groups fund specific
research projects, as do environmental organizations and foundations.

As marine ecosystems are fished more intensively, monitoring the
biological health of fish stocks and limiting their use become more
important. As more people enter the fisheries and total fishing power
increases, monitoring economic and social conditions also becomes
more critical. The 1996 MSFCMA amendment further intensified the
need for scientific monitoring and assessment with requirements to
define overfishing, reduce bycatch, account for essential fish habitat,
and minimize the impacts of fishery regulation on coastal communities.
But funding for these tasks has not kept pace with requirements. Wors-
ening the situation is an increase in the proportion of the National
Marine Fisheries Service budget earmarked by Congress for special
projects. Concerns are growing about whether research funding is adequate to the task.

On the other hand, there are questions about how much the public
should subsidize fisheries through government-funded research. Are the
costs of obtaining scientific information consistent with the benefits of its
production? As one scientist notes, "Research is expensive and the cost of
adequate research may be high in relation to the benefits from a fishery.
How much is it worth?" Who bears those costs and how might they be
lessened? With increasing frequency, people are suggesting that those
who benefit economically from the fishery—fishermen, processors, and

communities—should pay the cost of research required to support fishing activities.

The "user pays" idea of cost-recovery is expanding to many natural resources. But a number of fishery stakeholders remain reluctant to accept the idea, arguing that the fishing industry should not be responsible for covering the costs of scientific research agendas. They question the real need for, and worth of, the research. Some believe that moving toward limited entry and individual quota systems will lead commercial fleets to increase voluntary support of fisheries research, as discussed in chapter 4.[27] If fleets are smaller and are more directly vested in the fishery, they say, the industry may be more willing to invest in research. But others are concerned that this investment will lead to industry requests for a greater say in management on the theory that "user pays, user says."

Some states and industries have moved in the direction of cost recovery by establishing taxes on landed catch, as has occurred in the sea urchin fishery of California. But, in federal fisheries, the MSFCMA limits landings fees to administrative costs, and—in the case of the North Pacific's individual quota fisheries—to management and enforcement costs directly related to the quota programs. This leads one manager to argue that, even though science is very much related to the success of any management program, the law appears to excessively limit the use of the National Marine Fisheries Service's cost recovery authority to help fund scientific research. Should industry pay for scientific data collection and analysis? If higher fees are needed to support research, who should pay them and on what basis? Should small-boat fishermen pay the same fee as large-boat fishermen? What about recreational anglers, who pay license fees to fish in most state waters but generally not in federal waters? How much should the public be willing to pay through general funds in the U.S. treasury for the benefits of well-managed fisheries?

Cooperative Research

> We have done a bad job of incorporating what fishermen know into our scientific assessments. Fishery managers at the state and federal level are doing themselves a disservice by not working to incorporate the knowledge of fishermen.
>
> *Jerry E. Clark, National Fish and Wildlife Foundation*

> There doesn't need to be a black and white separation between science and the people on the water.
>
> *Bob Storrs, Unalaska Native Fisherman's Association*

> Industry participation in data collection can be a help, but the science must mean something.
>
> *Margaret F. Hayes, National Oceanic and Atmospheric Administration*

State and federal agencies employ scientists to run research vessels; sample and measure landings; collect, manage, and analyze data; and write reports for decisionmakers. These scientists, in addition to some academics, are the primary contributors of science to fishery management. But the traditional approach to the production of scientific information is being changed by two important trends. The first is greater cooperation among agencies. The second is cooperation between agencies and the fishing industry. Declines in fishing opportunities, as well as concerns over the quantity and quality of data used in management decisions and the long-term viability of fishing, have resulted in industry requests to play a more active role in conservation through participation in research.[28] In the words of a west coast fisherman, "The entire industry deeply desires to help gather better data. Our lives depend on it."

Expanding information needs and a pattern of insufficient funding suggest that data collected under fishing conditions will become more important to the scientific analyses used in fishery management in coming years. Fishermen have already begun to increase cooperation with research institutions and with the National Marine Fisheries Service. Some industry groups, such as the west coast trawl fleet, actively promote industry contribution to research through the use of fishing vessels as research platforms for government and academic scientists. Other fishermen conduct research projects on their own, experimenting with ways to reduce bycatch or avoid marine mammals. Still others collect fisheries and oceanographic data while at sea and distill their own observations and reflections about fish abundance, behavior, and oceanic conditions—described by one scientist as the "time-series of information in their heads."

The National Marine Fisheries Service has also become more willing and able to engage in such cooperative and collaborative projects.[29] A good example is the joint effort of the west coast groundfish trawl fleet and the National Marine Fisheries Service's Northwest Fisheries Science Center to design and carry out at-sea research. On the East Coast, the longline swordfish industry gathered data on bycatch and distribution, and, more recently, the surf clam and ocean quahog, lobster, scallop, and herring industries have cooperated with the National Marine Fisheries Service to carry out special studies.[30] The National Marine Fisheries Service also conducts a Marine Recreational Fisheries Statistics Survey to assess the frequency and type of motivators for recreational fishing. But, according to one representative, many recreational fishermen believe they are not adequately represented in the survey and want to foster better relations between the recreational fishing interests and the scientific community.

Research cooperation among other groups is also increasing. Fishermen and biologists at the University of Alaska are investigating the impact of trawling on the benthic ecosystem. A Texas program to enhance red drum was developed by a public utility, the Gulf Coast Conservation Association, the Texas Parks and Wildlife Department, and the U.S. Fish and Wildlife Service.[31] And the University of California is proposing to establish a Stock Assessment Center where students can be trained in fish stock assessment while performing such assessments for the state's Department of Fish and Game.[32]

What are the possibilities for members of the fishing community—commercial or recreational—to become full partners in research rather than merely responding to surveys or supplying research platforms? How can fishermen's individual, at-sea, "anecdotal" experiences be systematically incorporated into science-based management? From the perspective of many, "fishery-dependent" data should include this type of "anecdotal" information.

As one remarks, "It's very difficult for the fishing community to see things on the water and have this ignored by the scientists." It is also difficult for many in the fishing industry to accept the credibility of scientific analyses made by people who do not go out on the water with fishermen. "Information from the industry is underused and almost ignored, and that is a tragedy," one stakeholder remarks. And still another: "We are forgoing opportunities to learn about population changes and also opportunities to learn about fishermen behavior."

Despite strong support and an increasing trend toward joint research, collaborative efforts remain controversial and opposition is strong. Opponents have different reasons for their negative view of industry involvement in data collection and research. One scientist notes that the information communicated by fishers is often not at all verifiable and may often be contradicted by other fishers. Some argue that using the industry is a "poor boy approach" or that "the truth is not in 'em." Others believe that the inclination of the fishing industry to provide accurate information or to participate in special studies is limited. "Most fishermen are afraid that information they provide will be used against them, so this distorts the information that they give," says one. The challenge is "how to deal with the inclination of people to be dishonest when their ox is being gored," another adds.

Industry-generated data are "subject to tremendous deliberate and inadvertent biases," a government official notes, and is therefore worth little in the absence of fishing-vessel observers or other reliability checks. One fisherman agrees, indicating that log book data should be used only as an index of what is happening because, in an unobserved fishery, log books are not reliable in their detail.

Some west coast fishermen strongly disagree, arguing that logbooks, which are prescribed by law, constitute a huge source of data that are useable, but not in their current form. They suggest that, at present, paper logbook data are used sparingly, and with declining frequency, largely because state agencies are understaffed and underfunded. These fishermen are developing a simple standardized electronic logbook program that, if used consistently throughout the fleet, would not only vastly increase the usefulness of standard logbook information, making trip data and landing data readily available for scientific analysis, but would also, over time, provide fishers with the ability to better plan fishing strategies. Regarding the accuracy of such information, they question the alternatives, noting that, if the National Marine Fisheries Service lacks the resources to collect this information on its own, the fleet itself is the only alternative.

An environmentalist in the western Pacific indicates that researchers in that area are not as concerned that industry-generated data are suspect, but rather that the industry is reluctant to provide any information other than what is required in logbooks because of fear that it will become widely available and used by competitors. Several others argue that it is not the industry, but the National Marine Fisheries Service that is mistrustful and unwilling to cooperate. So how can accuracy and confidentiality be ensured? Even most proponents of collaborative research agree that information derived from experience must be balanced with information derived scientifically. "I believe in the participation of stakeholders in decisions which affect their lives, just as I believe the National Marine Fisheries Service should be scrutinizing such decisions on behalf of the public trust," one remarks.

Some stakeholders note the need to educate fishermen about science, saying that many fishermen fail to understand the nuances of surveys and random sampling. "There's a section of the industry that doesn't understand the purpose of a survey—that a survey is not designed to find fish, but to get an index of abundance," a fisherman says. An environmentalist makes a similar point: "There seems to be a common misunderstanding of the nature of science and random sampling. This needs to be explained. With scientific oversight, fishermen can be valuable in helping to get research done."

One example of this misunderstanding is provided by a scientist in the Northeast, who explains that resource surveys assess the status of the resource over the entire area of the distribution, whereas fishermen, enabled by very sophisticated technology, generally tend to "sample" the areas of high concentration. According to this scientist, fishermen in the Gulf of Maine recently claimed that there is no problem with the cod stock, based on the high catch rates they achieved within two or three relatively small areas. But scientifically, statistically designed surveys later concluded that these were virtually the only areas in which Gulf of Maine cod were occurring, as compared to their much wider historical distribution patterns.

Almost all fishery stakeholders—including industry representatives—agree that the perfect joint research scenario would include an observer program in every region. But all recognize the difficulty of establishing such programs given present funding levels. In the North Pacific, the groundfish fleet pays for on-board observers of their operations. This program is praised by an industry spokesperson, who indicates that the value of the data collected under the program rests in part on the scientific oversight from the National Marine Fisheries Service. Unfortunately, these data are not used to full advantage—a problem documented in other regions, like New England, as well.[33] Failing to use data provided by industry can result in industry reluctance to cooperate in providing the data. Many believe that those who cooperate should receive positive feedback, either through a rapid distribution of research results or through enhanced fishing opportunities.

Participating in research has the added value of increasing industry's stake in science-based fishery management, some note, and should foster good stewardship. Whatever it takes to create long-term commitment will make fishers excellent partners with incentives to "fish clean and to demand to have their views known," an environmentalist adds. A fishery scientist agrees: "The only way we can manage is to make the fishing industry an integral part of the system. But we have a long way to go both technically and institutionally to accomplish this." Finding ways to bring the experience and resources of people engaged in commercial and recreational fishing "to the table" as legitimate complements to scientific information will be met with some skepticism.

Conclusion

It would help if we could find some medium ground for the practical end of industry and the scientific end. Neither is complete alone.

Dave Krusa, F/V Restless, *Montauk, New York*

The science doesn't affect the agenda as much as it affects the packaging.

Michael L. Weber, freelance writer and researcher

Strengthening the scientific basis of management has widespread support. This can be done by improving the rate and frequency of biological stock assessments, applying ecological principles to fishery management, improving the collection and analysis of social and economic data, better incorporating uncertainty into fishery management decisions through the precautionary approach, improving the communication of scientific information, and developing a framework for industry collaboration in data collection and research. These are major challenges that will be exacerbated by an environment of increasing political pressure and litigation and limited resources. The collection of user fees to assist with the costs of these activities is one mechanism that should be considered.

In Their Own Words*

*These statements reflect the personal views of interview participants and do not necessarily represent the positions of the institutions or organizations with which they are associated.

Charlie Bergmann, Lund's Fisheries

We need better science, and this means getting the fishermen involved in the work. Good science includes recognizing and using so-called anecdotal information. It's very difficult for the fishing community to see things on the water and have this ignored by scientists. On the East Coast, scientists say the summer flounder are in terrible shape but we can't get rid of them, and they say we should be able to walk from here to Europe on the backs of mackerel but we can't find them. Recreationists see increased size limits and decreased bag limits, yet they see more fish in the water. So? We have to get more real time in our management, come to terms with the science we use, and how, and get stakeholders involved.

Eugene C. Bricklemyer Jr., President, Aquatic Resources Conservation Group

Science has not been used to my satisfaction in the past. We have not had enough science given the complexity of what we are trying to do. There has been too much localization, without recognizing the larger-scale interactions. There has also been too much compartmentalization of scientific disciplines, where the fish scientists don't talk with the marine mammal scientists, or with the other interacting components of the system. The major need is to look at the marine ecosystem holistically. We don't have an organization that views the oceans in total as a resource, and whose goal it is to keep the whole system healthy. It doesn't matter if you are maintaining the production of a target species, if you are harming the rest of the system. The goal should be the health of the ecosystem.

Barry Fisher, President, Midwater Trawlers Cooperative

There is nowhere near the amount of biological information necessary to answer the question of what is a sustainable fishery. You can't have a sustainable fishery for groundfish when fishermen are catching mixed species in their trawls and biologists are looking only at single species. The science is lousy— it's too little and of the wrong type. We need greater industry collaboration in research. Users should have a greater voice in providing data and experience to fisheries management. Fishermen's experience is the best source of data scientists can get. But the scientists mistrust us. They say that all fishermen are cheats.

Sonja Fordham, Fisheries Project Manager, Center for Marine Conservation

New England groundfish management has a long history of consistently bad decisions, continual delays in taking management action, and a reluctance to follow scientific advice. There has been a trend to always question the science when all else fails that, unfortunately, still exists despite disastrous consequences. I think that, if scientists are unable to assess every stock every year, managers should take a precautionary approach. We should account for uncertainty and incorporate buffers to ensure that we don't overfish. Fisheries science is a tricky business, and people train for many years to practice it. Managers should respect these experts and base decisions on their scientific advice.

Peter Fricke, Social Anthropologist, Office of Sustainable Fisheries, National Marine Fisheries Service

Social and economic information isn't properly used in fishery management. Where we have used this information as a way of gauging what type of impacts a regulation will have, decision-making has been better—much tighter. Social and economic information will never be a substitute for clear-cut decision-making, but it can show the limitations of various alternatives.

For example, banning fishing for bay scallops in North Carolina two weeks before Christmas would reduce the opportunity for those fishermen to have Christmas money and would limit the availability of seafood to make the scallop pies traditionally used in Christmas meals. Councils have to take the responsibility to properly include social impact analyses in fishery management plans. We need to enforce the rule that plan development teams must be interdisciplinary.

Burr Heneman, Marine Policy Consultant

Fisheries management operates under the myth of adequate information. Our system assumes that it's possible to obtain adequate information to make management decisions that allow us to fish right up to the limit of sustainability. In many fisheries, this may not be possible, certainly not with the resources we are willing to commit to do the science. The complexity of what we are trying to achieve, the difficulties of conducting research in marine environments, the reasonable limits on funding, and the results we have achieved all argue that we should consider another approach to management.

James E. Kirkley, School of Marine Science, Virginia Institute of Marine Science

Biological science plays a very important role in fishery management. Social and economic sciences, however, tend to be an afterthought or used as the holy-water blessing for regulatory decisions. In case after case, the National Marine Fisheries Service will support a regulatory regime that satisfies a given biological objective with little or no knowledge of the associated social and economic impacts. Even when these impacts are presented, they are given little weight in the decision process. The agency and councils simply follow the old school in which biologists or stock assessment scientists determine the regulations. The major problem, of course, is the lack of adequate data. The councils and the National Marine Fisheries Service simply do not collect adequate social and economic data nor do they have adequate programs in place to regularly conduct social and economic impact assessments.

Dave Krusa, F/V *Restless*, Montauk, New York

I would develop the means for people with proven track records in fisheries to work with scientists. These guys have a lot to say. Scientists need this information. For example, there is a line off Nova Scotia where all U.S. swordfish catches stop. The scientists came out with all kinds of biological explanations for this—separate stocks, etc. If they had only talked to fishermen they would have known that no one will say they caught fish on the other side of that line because of insurance reasons. Communication needs to go both ways between fishermen and scientists. If it did, the National Marine Fisheries Service would have fewer lawsuits and would know more about what is landed and about the situation in the fishery.

Terrance J. Quinn II, Professor, School of Fisheries and Ocean Sciences, University of Alaska

Economics and social science lack a single conceptual framework that everyone can agree to as the basis for analysis. I've noticed that biological models have more credence among the biological science community than the economic or social models do in their science community. There isn't a single economic model that would provide an overall framework for analysis. I'd like to see the equivalent economic modeling approach to biological modeling so that when we do the biological stock assessment we can see what will happen to the fishermen stock. Direct input from fishermen would be particularly welcome in economic modeling.

Andrew A. Rosenberg, Deputy Assistant Administrator for Fisheries, National Marine Fisheries Service

The fishing industry has always been and will continue to be a participant in data collection and analysis. Assessments depend on commercial landings information provided by the industry. There is a public misperception about the extent of the use of these data. The industry needs to be more involved in the analysis and data collection, to pay for more of it, and to improve the understanding that you suffer when you withhold information. We need to be able to link a specific result of management to the information provided by fishermen.

Cynthia Sarthou, Director, Gulf Restoration Network

Politics are controlling the system—it has nothing to do with what's best for the fisheries. Scientific information is only used to the extent that it fits what the council wants to do. They use it to support decisions they agree with but, if they're not in agreement, it's discarded. For example, in the Gulf, the same science that was used to justify the need for implementation of bycatch reduction devices due to highly overfished species was totally disregarded as a basis to set total allowable catch for red snapper. They said the stock assessment was totally ridiculous. We need to start listening to the scientists.

Courtland L. Smith, Professor of Anthropology, Oregon State University

The current process is too fish-oriented. There's not enough consideration of the human component. That's too narrow a view. It's better to think of what kinds of activities we want associated with the fisheries. I don't think the problem is a lack of social and economic information; it's how to use the information. We need to involve people that can deal with human conflicts and human relations. Biologists, although certainly receptive to social science information, are not trained to manage people or to put social science information to use. Most managers are concerned with protecting the fish, not with managing the users. If we don't change our orientation, we won't have fish or fishermen left to manage.

Lisa Speer, Co-director, Ocean Protection Initiative, Natural Resources Defense Council

We need to deal with how management measures will be made in the face of uncertainty. There will never be enough information. The question is what we will do in the interim. Too often, a lack of science results in fishing. I've been seeing more proposals for precautionary management but more work needs to be done in this area. In addition to improving the science base, we need a better way of improving the process of managing with gaps in scientific information. That has to be addressed.

Gary Stauffer, Director, Resource Assessment, Conservation, and Engineering Division, Alaska Fisheries Science Center, National Marine Fisheries Service

All issues faced by the councils have direct or indirect allocation implications, but the management system is not well designed to allocate among U.S. participants. The National Marine Fisheries Service isn't set up to effectively evaluate allocation decisions. Fisheries science hasn't figured out how to quantify nonbiological criteria to judge the benefits of alternative allocations. Poor knowledge makes fertile ground for political arguments. Allocation decisions are now devisive, bitter, and controversial. We need to tap into the science of decisionmaking for restructuring our management scheme to allocate scarce resources, including identification of new criteria, analytical models and data collection needs to evaluate allocation alternatives. The very narrow perspective from the 1950s and '60s of managing fishery resources by maximizing sustainable yield provides little to no guidance for allocation. We need to figure out new federal system for allocation from the prespective of Congress, the councils, and the National Marine Fisheries Service as the management organizations, decisionmakers, and information providers.

Notes

1. 16 U.S.C. 1851(a)(2).
2. 16 U.S.C. 1851(a)(1).
3. 16 U.S.C. 1853(a)(1)(A).
4. 50 CFR 600.345.
5. NMFS. 1994. *Guidelines and Principles for Social Impact Assessment.* NOAA Technical Memorandum NMFS-F/SPO-16.
6. Buck, E. H. 1995. *Social Aspects of Federal Fishery Management.* CRS Report for Congress, 95-553 ENR, 25 April 1995.
7. NMFS. 1998. *Report to Congress: Status of Fisheries of the United States.* U.S. Department of Commerce, NOAA, October.
8. NRC. 1998. *Improving Fish Stock Assessments.* Committee on Fish Stock Assessment Methods, Ocean Studies Board. National Academy Press, Washington, DC.
9. NRC. 1998. *Review of Northeast Fishery Stock Assessments.* Committee to Review Northeast Fishery Stock Assessments, Ocean Studies Board. National Academy Press, Washington, DC.
10. Ibid.
11. NRC: Improving Fish Stock Assessments. 1998, n. 8, 9.

12. Ecosystem Principles Advisory Panel. 1999. *Ecosystem-Based Fishery Management*. A Report to Congress as Mandated by the Sustainable Fisheries Act Amendments to the Magnuson-Stevens Fishery Conservation and Management Act, 1996.

13. Ibid.

14. NMFS. 1997. *Fisheries of the United States, 1997*. Current Fishery Statistics No. 9700. NOAA, Silver Spring, MD.

15. CEQ. 1997. *The National Environmental Policy Act: A Study of Its Effectiveness after Twenty-five Years*. CEQ, Executive Office of the President, January.

16. Rosenberg, A. A., M. J. Fogarty, M. P. Sissenwine, J. R. Beddington, and J. G. Shepherd. 1993. Achieving Sustainable Use of Renewable Resources. *Science* 262 (November 5): 828–29; Sissenwine, M. P., and A. A. Rosenberg. 1993. Marine Fisheries at a Critical Juncture. *Fisheries* 18(10):6–14.

17. NRC. 1998, n. 9.

18. Collins, C. H. 1994. *Beyond Denial: The Northeastern Fisheries Crisis—Causes, Ramifications, and Choices for the Future*. March 23. 72 pp. Unpublished manuscript.

19. Myers, R. A., J. A. Hutchings, and N. J. Barrowman. 1997. Why Do Fish Stocks Collapse? The Example of Cod in Atlantic Canada. *Ecological Applications* 7: 91–106; Roy, N. 1996. What Went Wrong and What Can We Learn from It? In *Fisheries and Uncertainty: A Precautionary Approach to Resource Management*, edited by D. V. Gordon and G. R. Munro, 15–26. University of Calgary Press, Calgary, AB.

20. Article 5(c) and Article 6.5, respectively.

21. NRC. 1998, n. 9.

22. NMFS. 1998. *Technical Guidance on the Use of Precautionary Approaches to Implementing National Standard 1 of the Magnuson-Stevens Fishery and Management Act*. NOAA Technical Memorandum NMFS-F/SPO-31, August.

23. *North Carolina Fisheries Association v. Daley*, No. 2:97cv339 (E.D. Va. Oct. 10, 1997).

24. Ibid.

25. *Southern Offshore Fishing Association v. Daley*, 995 F. Supp. 1411 (M.D. Fl. 1998).

26. Ibid.

27. Scott, A. 1993. Obstacles to Fishery Self-Government. *Marine Resource Economics* 8:187–99.

28. Sissenwine, M. 1998. Joint Science/Industry Research Benefits All. *Commercial Fisheries News*, (October): 5B, 7B.

29. Ibid.

30. *Commercial Fisheries News* October 1998, Special Section B.

31. Rutledge, W. P., and G. C. Matlock. 1986. Mariculture and fisheries management—A Future Cooperative Approach. In *Fish Culture in Fisheries Management*, edited by R. H. Stroud. Proceedings of a Symposium on the Role of Fish Culture in Fisheries Management, March 31–April 3, 1985,

Lake Ozark, MO, American Fisheries Society, Fish Culture and Fisheries Management Sections, Bethesda, MD.
32. Christopher Dewees, program director, Sea Grant Marine Advisory, University of California, Davis. Personal communication.
33. NRC. 1998, n. 9.

CHAPTER 8

Evaluating Fishery Performance

> All through the system there is no accountability [in that] no one
> is even asking questions. Congress just passes another law. Who's
> keeping score? Nobody!
>
> <div align="right">Barry Fisher, <i>Midwater Trawlers Cooperative</i></div>

In an environment of ever-increasing resource scarcity and evertightening regulations, fishery management is often driven by crisis. Management decisions tend to be reactive, rather than strategic actions based on long-term goals and objectives. For this reason, the resolution of one problem often leads to the generation of another, and decisionmakers continue to jump from one crisis to the next.

Fishery stakeholders from industry, environmental groups, government, and academia agree that routine assessment of the impacts and effectiveness of management decisions is critical to the success of fishery management. Even so, "the councils and [National Marine Fisheries Service] seldom evaluate whether or not their management decisions achieve the intended objectives," remarks one agency scientist. "They don't seem to have the time to learn what works and what doesn't from past decisions."

How does this failure to evaluate affect the development and implementation of fishery management plans? How does it affect management outcomes? Are fishery managers aware of the results—and unintended consequences—of their actions? "Being the one who has to bring management measures to the table, this is a big issue with me," a regional manager asserts. "It's embarrassing to propose new regulations when we can't even tell participants if the last measures worked or not. So let's evaluate and put the cards on the table."

Why Evaluate?

In 50 years we want an outcome that we intended, not a surprise.

Bob Storrs, Unalaska Native Fisherman's Association

The MSFCMA provides a broad goal for fishery management in the form of "greatest overall benefit to the nation." Ten national standards provide guidelines for meeting this goal (listed in chapter 1). These guidelines are important in terms of the broad oversight they provide. But they are not objectives. They are not measurable. Instead, it is up to the regional fishery management councils to determine specific objectives for the fisheries they manage, within the framework of the national standards and the overall goal of achieving the greatest overall benefit to the nation. Unfortunately, the vagueness—or sometimes sheer quantity alone—of goals and objectives specified in fishery management plans makes it difficult for managers to evaluate the success of specific management measures and their impact on fishery performance.

The Pacific Coast Groundfish Fishery Management Plan, as amended and implemented by the Pacific Fishery Management Council, provides one example of a plan with a large number of objectives that are difficult to quantify:[1]

Objective 1. Maintain an information flow on the status of the fishery and the fishery resources which allows for informed management decisions as the fishery occurs;

Objective 2. Adopt harvest specifications and management measures consistent with resource stewardship responsibilities for each groundfish species or species group;

Objective 3. For species or species groups which are below the level necessary to produce maximum sustainable yield, consider rebuilding the stock to the maximum sustainable yield level and, if necessary, develop a plan to rebuild the stock;

Objective 4. Where conservation problems have been identified for non-groundfish species and the best scientific information shows that the groundfish fishery has a direct impact on the ability of that species to maintain its long-term reproductive health, the Council may consider establishing management measures to control the impacts of groundfish fishing on those species. Management measures may be imposed on the groundfish fishery to reduce fishing mortality of a non-groundfish species for documented conservation reasons. The action will be designed to minimize disruption of the groundfish fishery, in so far as consistent with the goal to minimize the bycatch of non-groundfish species, and will not preclude achievement of a quota, harvest guideline, or allocation of groundfish, if any, unless such action is required by other applicable law;

Objective 5. Describe and identify essential fish habitat, adverse impact on essential fish habitat, and other actions to conserve and enhance essential fish habitat, and adopt management measures that minimize, to the extent practicable, adverse impacts from fishing on essential fish habitat;

Objective 6. Attempt to achieve the greatest possible net economic benefit to the nation from the managed fisheries;

Objective 7. Identify those sectors of the groundfish fishery for which it is beneficial to promote year-round marketing opportunities and establish management policies that extend those sectors fishing and marketing opportunities as long as practicable during the fishing year;

Objective 8. Gear restrictions to minimize the necessity for other management measures will be used whenever practicable;

Objective 9. Develop management measures and policies that foster and encourage full utilization (harvesting and processing) of the Pacific coast groundfish resources by domestic fisheries;

Objective 10. Recognizing the multispecies nature of the fishery and establish a concept of managing by species and gear or by groups of inter-related species;

Objective 11. Strive to reduce the economic incentives and regulatory measures that lead to wastage of fish. Also, develop management measures that minimize bycatch to the extent practicable and, to the extent that bycatch cannot be avoided, minimize the mortality of such bycatch. In addition, promote and support monitoring programs to improve estimates of total fishing-related mortality and bycatch, as well as those to improve other information necessary to determine the extent to which it is practicable to reduce bycatch and bycatch mortality;

Objective 12. Provide for foreign participation in the fishery, consistent with the other goals to take that portion of the

optimum yield not utilized by domestic fisheries while mini-
mizing conflict with domestic fisheries;

Objective 13. When conservation actions are necessary to protect a stock or stock assemblage, attempt to develop management measures that will affect users equitably;

Objective 14. Minimize gear conflicts among resource users;

Objective 15. When considering alternative management
measures to resolve an issue, choose the measure that best
accomplishes the change with the least disruption of current domestic fishing practices, marketing procedures, and
environment;

Objective 16. Avoid unnecessary adverse impacts on small
entities;

Objective 17. Consider the importance of groundfish resources
to fishing communities, provide for the sustained participation
of fishing communities, and minimize adverse economic
impacts on fishing communities to the extent practicable; and

Objective 18. Promote the safety of human life at sea.

If managers were to evaluate the performance of Pacific coast groundfish management in relation to these objectives, what criteria would
they use? How would they measure progress in meeting them?

These objectives were established by the Pacific Council to accomplish three stated management goals related to conservation, economics,
and utilization: (1) prevent overfishing by managing for appropriate
harvest levels and prevent any net loss of the habitat of living marine
resources; (2) maximize the value of the groundfish as a whole; and (3)
achieve the maximum biological yield of the overall groundfish fishery,
promote year-round availability of quality seafood to the consumer, and
promote recreational fishing opportunities. Even if the objectives established to accomplish these goals were directly measurable, is it possible
to achieve them—or the larger goals—simultaneously? Although the
goals have been listed in order of priority, what priority would the many
objectives be assigned in cases where not all can be met?

Contradictory goals and objectives are not unique to the Pacific coast
groundfish plan or the Pacific Council. Rather, they appear to be a common product of the consultative, participatory process that drives our
fishery management system. In developing fishery management plans,

attempts to accommodate the many, diverse interests in fisheries tend to result in the identification of incompatible objectives, such as those described previously. At present, the Pacific Council is undergoing a strategic planning exercise to develop a long-term vision for the groundfish fishery.[2] An important component of this activity is the reexamination of the stated goals and objectives for this fishery with respect to their practicality, consistency, and likelihood of achieving the council's long-term vision. Identification of a long-term vision is of little use without specific, compatible objectives that make progress toward achieving this vision measurable.

Consider, for example, the most recent National Oceanic and Atmospheric Administration Fisheries Strategic Plan,[3] which lists the following performance measures, committing the National Marine Fisheries Service, by May 2002, to

- increase the economic value derived by the nation from commercial and recreational fisheries;
- manage healthy fish stocks to achieve optimum yield;
- reduce the number of overcapitalized fisheries and mitigate the impacts of these reductions on fishing communities;
- minimize bycatch to the extent practicable and minimize the mortality of unavoidable bycatch; and
- implement the fisheries plan to achieve the goals of the interagency Recreational Fishery Resources Conservation Plan.

Although these objectives are in fact measurable, exactly how each will be measured is not apparent. While the increase in economic value derived from commercial and recreational fisheries may be relatively straightforward, how will a reduction in overcapitalization be measured? What criteria will be used to demonstrate how the economic impacts of capacity reduction on fishing communities have been curbed? This plan makes clear the wide variety of interests that must be balanced in achieving its objectives. What is not so evident is how these measures will indicate the agency's success in balancing those interests.

At present, failure to define specific, measurable objectives in fishery management plans leaves fishery managers vulnerable to political pressure, as they are forced to manage fishery resources for these competing, and often conflicting, goals. Concrete objectives would provide them a basis for responding to complaints about specific measures designed to accomplish those objectives and would also provide reference points against which to compare particular actions.

A recent report on the use of a trap certificate program to regulate Florida's spiny lobster fishery[4] provides a noteworthy example of the type of evaluation that is possible when a single, prestated, measurable objective is described—in this case, to reduce the total number of lobster traps. The evaluation shows that, five years after the program had been implemented, the number of legal traps operating in the commercial fishery had declined by 35 percent. The report also reviews the validity of early management expectations that the program would increase the yield per trap and maintain or increase overall catch levels.[5] It reveals that the average yield per trap had increased without affecting overall catch levels. Additional information on the resulting transfer of certificate ownership and use, administrative costs, and total revenues allowed scientists to include specific recommendations in the report. They urge managers to eliminate the surcharge on certain certificate sales and not to reallocate abandoned certificates during years of no reductions, to prevent the entry of nontrap gear groups in the fishery, and more. This guidance will enable managers to correct deficiencies and shape the program to provide further societal benefits. Required, regular evaluation of this type would help managers develop more specific evaluation of trade-offs—what we want from our fisheries over the long term and what we are willing, or not willing, to sacrifice to secure it.

Fisheries are complex and managers often cannot predict how the many and diverse components of each will be affected by specific management measures. Routine evaluation also would allow managers to better understand the consequences of their actions and adapt management as a result in a way that will increase the likelihood of meeting fishery objectives. Adaptive management, by instituting a learning capacity in the system, would allow fishery management to be more effective, efficient, and dynamic and enable managers to avoid repeat mistakes and make better decisions in the future.

A 1997 report by the Council on Environmental Quality, advocating a greater role for adaptive management in the National Environmental Policy Act process, states: "The old paradigm for environmental management was predict, mitigate, and implement. A new paradigm has emerged: predict, mitigate, implement, monitor, and adapt. The two latest threads—monitor and adapt—reflect the need to monitor the accuracy of predictions and allow enough flexibility in the process for midcourse corrections."[6]

Evaluation is the cornerstone of adaptive management, but it is not the only aspect of the fishery management system that presents managers with formidable challenges to managing adaptively. Embroiled in the nitty-gritty of allocations, setting acceptable catch levels, and other routine decisionmaking, council members have no space or time to examine their actions or the system retrospectively, let alone learn from the system in order to think strategically for the long term. Some we interviewed even question whether the councils are up to the task. They think strategic planning about fisheries in their regions would require councils and council system participants to come to agreement on future goals, a job that entails balancing many diverse interests.

A state fishery manager who participates in the council system believes that people are willing to work through issues, but that the process is not easy or swift. He notes that the pressure of decision-making generally does not allow managers to consider all the legitimate values that inform and drive goals. "Government can play a powerful role as a facilitator," he says, "to make it possible for communities to make their own decisions" and think in a longer-term fashion about structure, process, and other issues.

Whether the councils can take on such a job is questionable. Many observers believe that council members are not given sufficient training and background to set acceptable catch levels and allocate fishery resources among competing interests, much less take on evaluation or planning. Further, engaging stakeholders in such a process would require a different method for informing them about the choices, facts, and issues, as well as providing tools for their participation. The current system of council meeting as venue to pursue one's individual objectives would not be effective for collective long-range planning.

Can we construct opportunities for people to exchange information, debate, and come to agreement on general goals and strategies without allocations at stake? What can be done within the existing structure? Should advisory panels be given more authority and resources to develop goals and strategies? Are there opportunities for the councils to meet outside their normal meeting schedule, to share experiences from different regions, and to consider long-term objectives and strategic planning?

If councils are to have the space and time to do any long-term thinking—for example about ecosystem management or meeting new requirements to protect fish habitat—they need a different atmosphere than the pressure cooker of allocation, deadlines, disputes, and threatened litigation. But reaching agreement on evaluation criteria is the real challenge, according to one academic. "If we could agree on what criteria are, the gaps could be filled with some well thought out studies in a reasonable amount of time. The serious limiting factor now is the inability to actually agree on objectives."

Criteria and Objectives

> Once we've defined goals and objectives, then we can develop tools
> and standards for evaluation. Unfortunately, in most cases we have
> never articulated a vision of what we want our fisheries to look like
> in the future; so measuring success is problematic.
>
> *Scott Burns, World Wildlife Fund*

While there are differences among fishery stakeholders over which fishery management objectives are more important, the majority appears to
agree that some balance should be maintained between biological, economic, and social objectives. Almost all recognize that, in contrast to
the standard framework used to evaluate the status of biological objectives, the infrastructure necessary to systematically identify or evaluate
economic and social objectives is lacking. Perhaps for this reason, there
appears to be much more agreement on biological, as opposed to social
and economic, criteria and indicators.

A few stakeholders argue that a sophisticated evaluation system isn't
necessary, one scientist asserting, "We have adequate information to
evaluate fishery management performance. We know full well which
are doing well and which are not. We don't need a crystal ball or big
models." But the clear divergence of opinion voiced at council meetings
and in public hearings on fishery management plans belies the notion
that some things are obvious. It is not uncommon—even with appar-
ently collapsed fisheries—to hear testimony that stocks are healthy,
despite clear scientific indicators of depletion.

The biological health of a fish stock can be evidenced in the size and
age distribution of fish, the stock's geographic range, its spawning ability compared to spawning potential in an unfished state, the rate at
which it recovers from fishing, and its productivity over the long term.
Some stakeholders also suggest that ecological factors such as habitat
condition, system biodiversity, and quantity of bycatch and waste could
serve as indicators of biological health. But many note that there are virtually no baseline data against which to measure performance at the
ecosystem scale. Some believe that the ability to generate such baseline
information is an added benefit of marine protected areas as a fishery
management tool.

Fishery stakeholders most frequently choose sustained biological
production as the criterion against which to measure biological health.
But sustained at what level? While biological sustainability may be possible at a wide variety of stock levels, the MSFCMA now requires stocks
to be maintained at a level that will allow the production of maximum

sustainable yield. The law does not require that maximum sustainable yield be taken, so its intent to maintain the stocks' capability of producing maximum sustainable yield may be met while yields are kept deliberately low.

For example, a council may make an explicit decision to have a fishery for small, juvenile fish if smaller fish have a higher market value than larger, adult fish. In these cases, if the adult spawning biomass is properly protected, the ability of the stock to produce maximum sustainable yield will be ensured, although actual annual yield is much lower. Similar situations could arise for roe fisheries, sardine fisheries, little-neck clam fisheries, or others in which potential short-term biological yield is forgone in order to obtain greater social or economic benefits.

Such trade-offs are common and constitute a contentious and vexing part of fishery management. For example, while an economist may consider the catching of juvenile fish to have clear merit if it leads to greater overall economic benefits from the resource, others may consider such catches—either intentional or unintentional—to be unacceptable. This is one of the ways in which values are brought to bear on fishery management decisions. Taking fish for roe, fishing on spawning grounds, turning fish into fish meal, catching fish for recreation, large-scale fishing operations, the pursuit of local fisheries by fishermen outside the community, and myriad other fishing practices raise strong feelings among those interested in fishery resources, including our interview respondents.

The following comments exemplify two very different perspectives on management objectives. The first is taken from an industry representative's response to the question of acceptable trade-offs.

> There are other possible points the nation could go to, such as turning over the industry to large factory trawler operations, but these are very much suboptimal. While they may be efficient at catching large volumes of fish, they can easily lead to lobbying and pressures for pulse fishing. They wipe away the opportunities for communities to develop niche markets and products, they obliterate the good high-income jobs that fishermen love and replace them with low-wage industrial-style work, they strip community economic benefits and destroy social capital and seaport cultures, and they are generally very destructive of the ecosystem. Similarly, transferable quota schemes can lead to concentration of economic power, large vessels, pressures for lobbying for pulse fishing, an ethic of

private property versus community responsibility, and destruction of social capital and communities.

A contrasting view is taken from a discussion with an economist about the term "sustainable fisheries."

> The management of fish ought to be aimed long term at the maximization of economic return from the fishery. . . . A sustainable fishery would mean approaching a level of profitability sufficient to withstand the inevitable ups and downs of the market and stock abundance and therefore providing satisfactory, but never stable, return for the participants. I am concerned that when we start trying to maximize net returns then we slide from maximum earnings to protecting individual groups. To some, sustainability means sustaining everything in the position they're in now—I don't buy into this. All claimants got into the resource at different times for different reasons, and the idea that they all need protection is not supportable. . . . "Sustainable" means the best use of the resource. . . . Distribution of income, employment stability, etc. are secondary matters best dealt with if the fishery is biologically productive and economically profitable.

These are the different values that drive fishery decisionmaking. But baseline information to assess and reflect these values does not exist. Measurements of social capital, community cultures, the extent of lobbying for pulse fishing, or the conservation records of large- versus small-scale fishing industries do not exist. Fishery managers have no way of addressing, or validating, these concerns.

Better monitoring and evaluation measures would provide the fishery management system with a way to respond to such claims in a more objective manner. Perhaps the most frequent comment heard at council meetings is "you'll put us out of business." But the system does not monitor even basic measures of fishing business performance. Managers are rarely provided data on profit positions, turnover rates, employment, wages, and other characteristics of the labor force. The inadequacy of social scientific information leaves fishery managers with few options for assessing the social status of each fishery.

Given the U.S. experience of years of fishery decisionmaking focused primarily on biological objectives, the collection and incorporation of social data may appear to be an overwhelming—and perhaps unnecessary—task. Most fishery stakeholders we interviewed think not. They recognize that a fishery consists of more than just fish, noting that the people involved in fisheries and their social and economic health are integral components of a robust fishery. "The criteria for evaluation should include a look at how it affects people—a careful review of social and economic impacts," an industry sector representative asserts. "Managers should be required to know the entity they're managing before they regulate it."

Fishery stakeholders offer additional criteria that would serve as useful measures of industry status, including vessel sinkings, industry structure, fleet composition, value of vessels and licenses, vessel profitability, and value of capital in the fishery. Stakeholders also emphasize a need for indicators to determine the status of coastal fishing communities. They suggest using indicators such as population, age structure, commuting distances to work, number of serious crimes, educational attainment, ethnic population breakdown, and net migration and economic diversification.

Other criteria that would offer insight into the social and economic status of fisheries are geographic distribution of effort, distribution of wealth, equity and fairness, product supply, quality and price, consumer satisfaction, technological and economic efficiency, and overall return of commercial and recreational fisheries to the U.S. economy. Stakeholders also suggest the examination of nonmarket values such as the quality of recreational fishing experiences, aesthetic enjoyment of fish, and public satisfaction with the state of fishery resources. Management costs should be considered as well, some say, along with learning capacity, transaction costs, degree and nature of participation in the process, and enforcement.

Many of these data exist in some form, either at the National Marine Fisheries Service or elsewhere, such as the Census Bureau or independent research institutions. But, without a standard conceptual framework for routine and systematic data collection and analysis, and with vague objectives that are hard to measure, it is virtually impossible to apply this information uniformly to fishery management decisions. As a biologist notes, "On the social and economic side, I've never seen a plan put down something that's articulated in a measurable way."

A few people suggest that fishery stakeholders' and, in particular, fishers' views of management can be used as an indicator of performance. But others disagree. One argues, "We should not judge the performance of fishery management by the peace and quiet of the industry. To the extent that they are satisfying the industry, managers are not doing their jobs."

A participant in the North Pacific region notes that an evaluation of fishery management measures should include an examination of the associated enforcement measures, the efforts taken by enforcement agencies to carry them out, and the level of support by resource users for the management measures. He suggests that such an analysis will provide managers with valuable information on the return on enforcement expenditures or the opportunity costs for having failed to make those expenditures.

Many people we interviewed propose that the individual performance of fishery managers be evaluated as well, indicating that evaluations would institute accountability for management decisions. "Are fishery managers protecting the productivity of stocks for which they are responsible?" one long-time observer questions. "If stocks are declining for reasons related to management," he says, "they need to correct the situation." And a former administrator declares: "The council has the authority and, indeed, the responsibility to provide effective leadership. Some mechanism should be set up to suggest, urge, monitor, and evaluate council performance."

What Is Evaluated?

> [The National Marine Fisheries Service] has a standard format for biological evaluations, but they tend to ignore economic and social evaluation.
>
> Ray Hilborn, University of Washington

The MSFCMA and implementing regulations promote a fair amount of evaluation through specific provisions in the fishery management process. But there is a common perception among stakeholders that fisheries and associated fishery management plans are not subject to sufficient evaluation. This perception may arise from the emphasis that many place on the economic and social aspects of fishery management because, in actuality, there are a number of regular, formal, biological analyses that are conducted at both the national and the regional levels.

The MSFCMA, as amended in 1996, mandates that the National Marine Fisheries Service report to Congress and the councils annually on the status and condition of fish stocks, identifying fisheries that are overfished or approaching an overfished condition (estimated to become overfished within two years).[7] To make possible this determination, the act requires that each fishery management plan specify objective and measurable overfishing criteria. The mandate also requires that the secretary of commerce review fishery management plans or regulations developed for fisheries that meet these criteria at least every two years to determine if adequate progress is being made to end overfishing and rebuild fish stocks.[8]

In addition to these annual reports, the National Marine Fisheries Service produces the *Our Living Oceans* series, which includes periodic reports on the biological status of fish stocks. Although the first publication was produced before the Government Performance and Results Act of 1993 required government agencies to have a performance-based strategic plan with measurable objectives, *Our Living Oceans* now represents one aspect of meeting this requirement. The other aspect is the National Oceanic and Atmospheric Administration's Fisheries Strategic Plan, which outlines national fishery objectives. The *Our Living Oceans* series is intended to link the strategic plan to both fishery performance and results. Specifically, the series thoroughly reviews the biological status of U.S. marine fishery resources, calling attention to particular management problems that face the National Marine Fisheries Service.[9]

Both the National Marine Fisheries Service and the councils produce periodic regional assessments of the biological status of managed stocks as well. Some are discretionary, such as *The Status of Fishery Resources off the Northeastern United States*,[10] others mandatory, such as Stock Assessment and Fishery Evaluation reports, which must be prepared—and updated annually—for each stock managed under the councils' purview. The latter are intended to provide the councils with the information necessary to determine annual catch levels, to document significant trends or changes in the fisheries over time, and to assess the performance of state and federal fishery management programs.

Each Stock Assessment and Fishery Evaluation report is required to summarize the social and economic condition of the fishery's recreational, commercial, and processing sectors, as well as the most recent biological status of the fishery.[11] But these reports usually include social and economic information at only a rudimentary level. Apart from this information, no social science evaluations are routinely performed at the regional or national scale. Although the councils periodically commission reports on the social and economic state of fisheries, these efforts are generally neither systematic nor widespread.[12]

In December 1996, as a part of the *Our Living Oceans* series, the National Marine Fisheries Service published a report on the economic status of U.S. fisheries.[13] Intended to describe the economic health of the fisheries, to provide baseline assessment data from which to measure progress, and to identify major data needs, the report focuses principally on the commercial sector but also describes the recreational, commercial processing, international trade, and retail sectors. Despite a suggestion in the foreword that the publication is an effort that will be periodically repeated, the National Marine Fisheries Service reports that a sequel is not likely to be published until at least 2001.[14]

The emphasis here on social and economic objectives is not meant to diminish the importance of biological objectives. Rather, it is meant to show that a more structured evaluation process is required. The social effects of fishery management will continue to increase in importance as fisheries move to a new era of conservation and sustainability. Management needs a way to better assess these components in a structured, organized manner and recognize their impact on the process in order to make the transition successful.

Social science evaluation based on explicit objectives would provide the management system with a structured way in which to analyze the impacts of fishery management decisions and clearly identify trade-offs. As one academic remarks, "Overall, fishery management should be done to produce the largest possible net benefits to the public. Then, if choices are made to the effect that net benefits are not maximized, we should recognize that we are giving up something to make those choices. In practice, we do it the other way around."

In some cases, at least, this would avoid situations in which biology is torqued to meet unstated social goals. That many major fisheries were depleted in spite of biological advice and warnings indicates that social factors played a primary role in allowing overfishing to occur. "You can always learn more by investing more in biological investigations but, by comparison, our understanding of the human organism is completely rudimentary," an environmentalist explains. "We need more understanding of the human element in the fisheries management matrix, because that's something we can do something about, in contrast with fish, or currents, or primary productivity." Improved understanding of human behavior will also help managers and scientists develop incentive-based fishery management tools to promote conservation.

Where Do We Go from Here?

> There is not adequate monitoring, there is not adequate data, and there is not an adequate protocol.
>
> *Michael K. Orbach, Duke University*

The majority of industry representatives, environmentalists, managers, and scientists we interviewed agree that a regular, formal system of evaluation is critical to achieving healthy fisheries and effective fishery management. Given that agreement, what can be done? Diverse views about goals and objectives exist, and the crisis-driven, reactionary basis of legislative provisions and fishery regulations is a fact of life. The major gaps in baseline information are easily identified. Constraints on time and budgets are well known. So where do we go from here?

While its legislative mandates are clear, the National Marine Fisheries Service asserts that it does not have the human or fiscal resources to adequately evaluate all aspects—especially social—of fishery performance. Even if the agency elevated the importance of such activities, congressional earmarking of the National Marine Fisheries Service budget for specific purposes does not allow large redirections of effort. Unless Congress begins to back legislative requirements with budget allocations, such activities will likely remain neglected.

If, as many of those we interviewed suggest, more structured, regular, and comprehensive evaluation of fishery performance is desirable, and the National Marine Fisheries Service is not in a position to take this on, how else might it be accomplished? One suggests that some information and human resource needs could be met by increasing cooperative efforts with universities, nongovernmental organizations, and the fishing industry. There is also a widespread belief that either comanagement or rights-based management would stimulate the fishing community both to produce the data required for evaluation and to secure the analysis of that data.

Evaluation requires data, criteria, and analysis. Councils would be forced to consider all three if requirements to do forward planning were in place. Some stakeholders suggest that requiring councils to include measurable objectives for a variety of criteria, a definite timeline, and periodic evaluation in the fishery management plan process would have the same effect. Training and orientation in strategic thinking could help council members and staff with such requirements.

Many fishery stakeholders question whether the above is even possible, in light of the overwhelming information needs and the reality of limited financial and human resources. Although additional data are indeed a serious need, the fact that data available now are not being used effectively in decisionmaking cannot be ignored. Better analysis of existing data could be done.

A conservationist proposes that environmental groups, by creating expectations and reporting on the progress of fishery management in meeting these expectations, can play an important role in fisheries evaluation. A number of others suggest the use of independent evaluation committees as an objective and cost-saving device.

Conclusion

> We've lost the opportunity to make rapid changes or manage
> adaptively because we've not been evaluating the effects of our
> management.
>
> *Graciela Garcia-Moliner, Caribbean Fishery Management Council*

It is important to develop formal evaluation plans and routinely evaluate the performance of fishery management using concrete, measurable
objectives. Evaluation criteria should be part of all fishery management
plans. Systematic evaluation will help managers to determine the effectiveness and impacts of fishery management measures, recognize associated trade-offs, and adapt management to long-term goals. Despite
strong support for a formal evaluation process, supporters and critics of
the system alike point out that the councils face severe challenges in
being able to evaluate and learn from past actions. Learning about fishery management and the effectiveness of management measures chosen
will require resources and time, both of which are in short supply for
councils burdened by expanding workloads. However expensive evaluation might seem, forgoing evaluation and the opportunity to adapt
management accordingly may come at an even greater expense.

In Their Own Words*

*These statements reflect the personal views of interview participants and do not nec-
essarily represent the positions of the institutions or organizations with which they are
associated.

Linda Behnken, Executive Director,
Alaska Longline Fishermen's Association

Evaluation is timely and difficult to do effectively but it's necessary to know if we are moving in the right direction with the
decisions that we make. The councils should ask themselves
the questions, Was what we did effective over the last five
years? Did we achieve our goals? Did our solution solve the
problem that we set out to solve? The councils don't do
enough to maintain or follow up on decisions that they make,
unless one of these decisions is followed by another problem.
They could better evaluate the effectiveness of their decisions
and the achievement of their goals with better baseline data
and, also, if the evaluation process was built into the council
system process.

Jim Crutchfield, Professor Emeritus, Economics and Marine Affairs, University of Washington

To evaluate performance we should ask, Are fishery managers protecting the productivity of stocks for which they are responsible? If stocks are declining for reasons related to management, they need to correct the situation. Are they considering the economic impacts of actions on various groups? These need to be assessed in a fashion subject to peer review so that they can be done correctly. Overall, fishery management should be done to produce the largest possible net benefits to the public. Then, if choices are made to the effect that net benefits are not maximized, we should recognize that we are giving up something to make those choices. In practice, we do it the other way around. Evaluation is possible to do, and it needs to be done, but it needs an independent overseer. We need knowledgeable people to look at it carefully.

Rod Fujita, Marine Ecologist, Environmental Defense Fund

Evaluation of marine environmental policy should go beyond economic indicators such as profit margin, total revenue generated, and maximum sustainable yield targets. Ecological indicators of the status of marine systems are very important. We need to develop a way to quantify ecological factors—a baseline. Very few control areas have been established to serve as a reference condition. If scientific research is carried out in areas that are heavily fished, the results, such as effects on community composition, economic structure, and function are difficult to infer. Without a control, fish abundance and distribution are hard to interpret because humans have taken substantial components away. Marine reserves could help to establish control areas, serve as measurable indicators of ecosystem health, and set performance standards, as well as protect marine ecosystems and probably enhance fishery yields.

Timothy Hennessey, Professor of Political Science and Professor of Marine Affairs, University of Rhode Island

Obviously, if the fishery is in total collapse, it's not good. Beyond that, we should look at whether we are maintaining reasonable fish stocks and at the same time have equity and fairness in the management system. We should evaluate to

what extent people are learning, as similar situations develop
in one fishery after another. We should look at the huge transaction costs of decisionmaking. We should do the same for
enforcement, and the same with science. We could look at the
condition of the fishery in relation to the management system—it can't all be natural perturbations. We should practice
ongoing evaluation. Every two to three years we could get
data on management, enforcement, and science and really try
to get at these questions in a more rigorous way.

**Brian J. Rothschild, Director, Center for Marine Science and
Technology, University of Massachusetts, Dartmouth**

Who's doing any evaluation? Was the buy-back a success? The
National Marine Fisheries Service should not be evaluating performance, academics should not be evaluating performance,
and consultants should not be doing it. A non–National Marine
Fisheries Service governmental body should institutionalize
the evaluation of management performance. A Civil Aeronautics Board–type structure would provide for continuous evaluation and an institutionalization of the process. I'd suggest that
an oversight body comparable to the Civil Aeronautics Board be
developed to maintain national oversight and contribute to
checks and balances. A body like that would look over the
shoulder of the National Marine Fisheries Service much like
the Civil Aeronautics Board/Federal Aviation Administration
interaction.

Notes

1. Pacific Fishery Management Council. 1998. Amendment 11 to the Fishery Management Plan for Pacific Coast Groundfish. Pacific Fishery Management Council, Portland, OR.
2. June 1999. *Pacific Council News,* 23(3). A publication of the Pacific Fishery Management Council.
3. NOAA. NOAA Fisheries Strategic Plan (May 1997). U.S. Department of
   Commerce, Washington, DC.
4. Milon, J. W., S. L. Larkin, D. J. Lee, K. J. Quigley, and C. M. Adams. 1998.
   The Performance of Florida's Spiny Lobster Trap Certificate Program. Florida
   Sea Grant Program, University of Florida, Gainesville.
5. Florida Statute 370.142(1).
6. CEQ. 1997. *The National Environmental Policy Act: A Study of Its Effectiveness after Twenty-five Years.* CEQ, Executive Office of the President, January.

7. NMFS. 1998. *Report to Congress: Status of Fisheries of the United States.* U.S. Department of Commerce, NOAA, October.
8. 16 U.S.C. 1854(e).
9. NOAA. 1997. *Our Living Oceans. Report on the Status of U.S. Living Marine Resources, 1997.* NOAA.
10. NMFS. 1998. *The Status of Fishery Resources off the Northeastern United States.* NMFS, Northeast Fisheries Science Center, Woods Hole, MA.
11. 50 CFR 600.315.
12. See the 1994 NPFMC Faces of the Fisheries series for a good example of a social science assessment prepared for the North Pacific Fishery Management Council by an independent contractor and funded by NOAA.
13. NMFS. 1996. *Our Living Oceans. The Economic Status of U.S. Fisheries, 1996.* NOAA Technical Memorandum NMFS-F/SPO-22, December.
14. Personal communication. NMFS Office of Science and Technology.

PART IV

The Future

We need to get beyond the perception that fishermen are all out there to rape the resource. We need to get beyond the perception that all the oceans and everything in them are dying. We need to get beyond the perception that all scientists are sitting in their laboratories trying to figure out ways to screw fishermen.

Rod Moore,
West Coast Seafood Processors Association

CHAPTER 9

Looking Ahead

With regard to public policy: the car is stalled, we're sitting on the train tracks, the train is coming, we can hear the whistle, and we're arguing over what station to put the radio on.

Richard Young, F/V City of Eureka, *Crescent City, California*

For such a resource, you'd like to see more of an end point than just "let's make a deal and divvy things up."

Mike Nussman, American Sportfishing Association

The first two sections of this book present a range of fishery management issues. They summarize the findings of the published literature and the thoughts and perspectives of stakeholders on stock status, management history, productivity, ownership, management, incentives, scientific information, and evaluation. People in industry, environmental organizations, government, and academia voice diverse opinions on these issues.

From the diversity, a number of common themes emerge related to learning, integration, expectations, diversity, and transition. The themes cut across the issues that form the subjects of the previous chapters, as well as across interest groups, regions, and fisheries. They contain within them the policy choices that decisionmakers and fishery stakeholders face in shaping the future of U.S. marine fisheries. How can we best learn from our experience and integrate complex management objectives? How can we create expectations for stewardship, maintain the diversity of fisheries, and still make the transition to sustainability?

These choices are not easy, particularly when much of what Americans value about fishing are its history and tradition—which will inevitably come in conflict with the future realities of fishing. Will we learn from this history, or cling so steadfastly to tradition that we are doomed to repeat it?

Learning and Adapting

The "tragedy of the commons" is like the weather. Everybody talks about it but no one does anything about it.

Jim Gilmore, At-sea Processors Association

There are definitely some creative thinkers in the agency [National Marine Fisheries Service], but the process and the act [MSFCMA], in a sense, lock them into a reactive mode.

Alison Rieser, University of Maine

How do we learn from history? Or do we? A system to evaluate management performance and learn from past experience is critical to making progress in any endeavor. Unfortunately, such a system has generally been absent from natural resource management in the United States, including the management of marine fisheries. The need for two areas of learning surfaced in the previous chapters.

First, fishery stakeholders need to learn how to participate more effectively in the fishery management system, not only as members of councils, committees, and panels, but also as public owners of fishery resources. If the people who make decisions about the future of fishing are to learn to do this job well, resources have to be made available to foster informed participation and good decisionmaking. Opportunities for professional development, education, training, leadership development, and public information must be provided by the councils, National Marine Fisheries Service, and Congress.

Second, fishery managers must learn from the evaluation of prior decisions and models. Decisionmaking can be improved by routine evaluation of the effectiveness of fishery regulations. If decisionmakers are to learn from past experience, they need authority, time, and resources to evaluate the effectiveness of management measures so they can adapt. Evaluation is needed throughout the fishery management process. Better use of the evaluation mechanisms established in the National Environmental Policy Act, regulatory impact analysis, and economic analysis to assess the potential effects of proposed measures would counter the current trend of legal challenges to plans. Structured evaluation—of both biological and social assumptions—would provide a rational tool for adjustment, rather than reaction to stakeholder pressure when measures cause hardship. Endpoint evaluation would review performance to ask: did the measures do what they were supposed to in the time allowed?

Evaluating the performance of fishery management decisions can be useful to other marine activities as well, helping decisionmakers to avoid repeating mistakes of the past. For example, although marine aquaculture is already an important part of world fisheries, its development is in early stages in U.S. marine waters. We cannot afford to make the same mistakes in aquaculture that were made in fishery management. We can learn from past mistakes in managing capture fisheries and not repeat them as the U.S. marine aquaculture industry develops. Learning from our experience in capture fishery management, we can anticipate the harmful effects of subsidies, the problems created by unclear ownership, the need for planning, and the costs of avoiding the inevitable allocation questions. We can, before conflicts arise, plan for how aquaculture and capture fisheries will be integrated, how regulatory responsibilities will be distributed, and how the risks and trade-offs of environmental effects will be assessed.

What can fishery managers learn from other natural resource–based businesses and public resource regimes? What are the lessons from forestry, mining, or agriculture that might inform fishery management? Learning from the past and from the experience of others is key to making a transition to the future. Another key is to integrate what we learn—across regions, disciplines, and sectors—and to integrate complex management objectives.

Integrating Management Objectives

> Management by crisis is the way we are doing things. When a crisis comes your opportunities are limited; you have few opportunities to change.
>
> *Francis Christy, IMARIBA*

Over the past 22 years, the regional fishery management councils have faced increased legal and management responsibility and rising information needs. This has coincided, in many regions of the country, with the strain of having to allocate smaller amounts of fish among the same, or increasing, number of users. The problems that have arisen as a result of these management complications are numerous and have been outlined in previous chapters. They are indicated by excess capacity, overfished stocks, and low economic returns in many fisheries, and they reflect a strong need for integration over several spheres.

The MSFCMA should be reviewed in combination with other important marine statutes, such as the Marine Mammal Protection Act, Endangered Species Act, Coastal Zone Management Act, and Marin Sanctuaries Act, to examine how to integrate the entire system in a way that improves coordination, reduces conflicts, and moves toward ecosystem-based management.

The adequacy of funding to support the research and management responsibilities of the system should also be assessed. At present, funding appears to be seriously inadequate, and it is unlikely that future appropriations will reverse this trend. Fishery managers and administrators should review, prioritize, and integrate existing requirements. New requirements introduced by Congress should be accompanied by adequate funding and staffing to do the required research and management. Routinely evaluating the cost-effectiveness of alternative management and research approaches will help to stretch limited funds.

Ecosystem-based management is another important area needing integration. Recent MSFCMA provisions and implementing guidelines have begun to direct U.S. marine fishery management toward an ecosystem-based approach. It is important that the effects of traditional fishery management tools, such as minimum size limits, are considered as fishery management moves in this direction and that the use of these more traditional tools is integrated and coordinated with that of newer, ecosystem-based tools, such as marine protected areas, in a way that works to improve fishery management at the ecosystem scale.

How can integration across disciplines, regions, agencies, and jurisdictions be encouraged? Integration requires opening up the process to new information and providing mechanisms to do something with it. Of course, if you don't know where you are going, any road will get you there. Integration is a process for getting somewhere. Our expectations define where we want that to be.

Creating Expectations for Stewardship

> The Magnuson-Stevens Act wanted to be something to everyone.
>
> *Gil Radonski, formerly Sport Fishing Institute*

> We need to agree to a policy about risk and uncertainty and to decide on an acceptable mechanism for measuring risk and making trade-offs between the commercial value of fish harvests versus the "environmental" value of forgone harvests.
>
> *David Sampson, Oregon State University*

How can public ownership best be reconciled with fishery stewardship? How do we create consistent expectations about what is needed for stewardship and long-term sustainability? Is fishing a privilege or a right? What is the "greatest overall benefit to the nation"? These questions present fundamental choices about how fisheries will be managed in the United States.

As illustrated in the previous chapters, fishery stakeholders share widely differing views about many aspects of U.S. marine fisheries and fishery management. These differences in views reflect differing expectations. Some expect to be able to continue fishing and making a living, some expect to be run out of business, some expect that coastal communities will be preserved, some expect recreational fishing to be given more weight, some expect that the government will conserve the fish, and some remain unclear about what to expect. The region, the sector, the fishery, the gear group—each has expectations about what fishery management decisions should achieve, whether related to national policy or local allocations. It is when these interests collide, or when special interests bump up against the national interest, that conflict arises. That conflict is even more severe when we approach fishery management with different expectations.

Management and allocation of scarce resources cannot be all things to all people. By attempting to do so, the MSFCMA has created unreasonable expectations. Those who make policy choices must address these expectations, even the difficult ones, so that future expectations will be both shared, and more realistic. If Congress is to decide between a short- and long-term outlook for American fisheries, it must reexamine the goals and expectations set out in the act.

The incentives and disincentives in the fishery management system must also be examined. Management needs to be redesigned to promote behavior that achieves the current goals of fishery management, such as capacity reduction, minimization of bycatch, and the protection of essential fish habitat. Rights of access must be clarified. Illegal fishing activities must be penalized fairly and appropriately. A system of rewards is also needed to recognize responsible participation and stewardship. The use of incentive-based management tools will help managers to focus expectations on the long-term benefits provided by sustainable fisheries.

Maintaining Fishery Diversity

> There is no such thing as the "average fisherman."
>
> Ray Bogan, United Boatmen of New Jersey and New York

The previous chapters have served to remind us of the richness and diversity of U.S. fisheries. That diversity is expressed in recreational and commercial uses, in scales of operation, target species, gear types, ethnic groups, and regions. The regional fishery management council system has fostered that diversity, as the original FCMA intended. One cherished expectation of those involved in American fisheries is that their variety and differences will continue.

Diversity is a characteristic of U.S. fisheries that almost all would want maintained. The challenge for making the current transition is how to balance the wide variety of interests while rebuilding stocks, protecting fish habitat, and reducing fishing capacity. All these have implications for allocation, and allocation means winners and losers. Some are concerned that diversity in the fisheries may be reduced or lost.

Maintaining diversity is related to the choices of management objectives and management process. Managers are faced with difficult questions without a lot of guidance or information. How should long-term goals and objectives of fishery management be defined? What is the proper balance between utilization and conservation? What do we want fleets to look like? Do we want to sustain fishing communities? If so, what does this mean? What are the social and economic consequences of downsizing? How important are future generations compared to the present?

Preserving traditions of diversity and opportunities for a variety of voices to be heard demands long-range planning and stabilization of management. The crisis mode that accompanies many catch reductions and allocation disputes does not foster the kind of long-term thinking that enables consideration of alternatives for preserving diversity or articulation of a future vision for fisheries and fishing communities. In the pressure cooker environment of council meetings, the struggle for survival today outstrips any other consideration and chokes off opportunities to think about what fishing might look like tomorrow.

It may be possible to salvage the best elements of the council system—decentralized process, regional expertise, public participation, hands-on advice—but draw upon lessons from other models to improve the system. The checks on special interests must be strengthened to maintain a balance between special interests and the national public interest.

Making the Transition

> There was a lot of turmoil at the beginning that still hasn't gone away. Many of the impacts of early council choices are impossible to reverse.
>
> *Bill Gordon, formerly National Marine Fisheries Service*

> We are doing most things right in fishery management. We need
> resources to do the job, and we need to stick it out.
>
> *Michael K. Orbach, Duke University*

The two decades of fishery management under the regional fishery management councils have resulted in many changes. The future holds no likelihood that things will go back to the way they once were. Fisheries have developed to the point of overcapacity. Many fish stocks are stressed. The councils and the National Marine Fisheries Service are facing new realities, new mandates, and new challenges. The future of fisheries will not mimic the past. The pressures of increasing coastal populations, environmental stresses, and the need to manage at the ecosystem scale all mean that future fisheries and their management will be quite different from what we have known.

The globalization of markets, advances in processing and information technology, and increasing authority and responsibility in international or multinational regional bodies have made fishing an international business. The FCMA passage in 1976 was an exercise in claiming the U.S. fishery conservation zone for American fishermen, but the international interconnections of the twenty-first century demand that we also remain aware and responsive to international conservation concerns.

How will we retool for a more connected world? The legacy of past practices is with us in overfished stocks and overcapitalized fleets. A successful transition to sustainability will only be made possible if fishery conservation is strengthened and fishing capacity reduced. The transition will also require the planning and definition of long-term management goals and objectives. What do we want fisheries to look like 20 years from now? How can management most effectively address ecosystem concerns, rebuild stocks, provide appropriate incentives, and restore economic productivity to the fisheries? Who will pay the costs and enjoy the benefits of this transition?

It is important that policymakers recognize that they must make tough decisions to reduce capacity in the fisheries and rebuild stocks, even though the effects of those decisions will inflict pain on some of their constituents. It is appropriate for government to help fishing communities deal with transition, including education, economic assistance, and other measures to enable people to help shape their fishing future.

A successful transition also requires anticipating the emergence of new interests that may affect the future of fisheries, for example, aquaculture. The U.S. marine aquaculture industry is in its infancy but offers potential for growth in food production, wild fish supplementation, bait production, aquarium species production, and pharmaceuticals. How will marine aquaculture interact with marine capture fisheries? How will the two business cultures—the farming business of aquaculture and the hunting business of capture fishing—interact? Fishing and farming are not the same thing. But the operations of the two types of businesses may compete for ocean space as well as in the international seafood market. On the other hand, there may be little interaction on either front. The point for the transition is to anticipate emerging non-traditional activities with which fisheries may interact and for which fishery management will need to plan.

Writers on the topic of transition describe three phases to the process: (1) an ending and (2) a period of confusion and distress, which leads to (3) a new beginning. Clearly U.S. fisheries and fishery management have reached the end of the frontier. And the array of complaints, criticism, and points of view has defined a period of confusion and distress for some time. Will the distress lead to a new beginning? Or just more distress? How can this transition be made as painless as possible for those left behind?

It is important to recognize the ending, and see it as an opportunity as well as a loss. During the middle period, stakeholders and decision-makers need to use the time to shape a new idea for the future. A transition will be successful in its third, or new beginning, period only if we set new priorities. Those priorities must come from facing the policy choices that flow from learning and evaluating past experience and policies; integrating across legal, geographic, disciplinary, and sectoral boundaries; revising our expectations; and recognizing our diversity.

Major Policy Choices
and Reauthorization Issues

> The reauthorization should focus on better defining the goals and objectives of fisheries management, particularly with respect to society and not just the fish.
>
> James E. Kirkley, Virginia Institute of Marine Sciences

> The MSFCMA is now extraordinarily complicated. It is festooned with bells and whistles from special interest groups and needs to be simplified.
>
> William T. Burke, University of Washington

On the basis of interviews with stakeholders, review of the literature, and our own experience, we have identified a set of key issues that need resolution; some are characterized by widely disparate views with respect to how they should be resolved.

- **Define "Greatest Overall Benefit to the Nation"**: Stakeholders and policymakers must define what is meant by "greatest overall benefit to the Nation." Owners of fishery resources are all the people in the nation, so the discussion must include what is in the national, public interest, as well as the interests of user groups. Instead of operating with the idea that we can maximize an unlimited number of objectives, trade-offs should be presented and each addressed explicitly. This includes examining whether it is possible to achieve maximum benefits from every fish stock at the same time. It also includes setting management goals for the long term. A national strategy should be developed on which to base specific regional operating plans.
- **Develop a Strategy to Reduce Fishing Capacity**: Finding effective ways to reduce commercial fishing capacity to a level compatible with sustainable fishing is an urgent need. A number of actions should be considered, including eliminating subsidies, examining options for federally sponsored buybacks, and adopting systems of tradable fishing rights in which fishermen buy out excess capacity. The issue of capacity reduction in recreational fisheries in which the objective is to provide recreational fishing experiences is not as easily addressed, requiring careful scrutiny to determine how best to match capacity with the objectives of recreational fisheries. Because such actions will be painful for some participants, the federal government should be prepared to assist by easing the economic pain and providing education about new realities and alternatives. Government programs such as Sea-Grant and the Small Business Administration have the necessary infrastructure.
- **Resolve Access to Fishery Resources**: In light of the excess fishing capacity that currently plagues U.S. marine fisheries, it is imperative that policymakers limit access to commercial fisheries. As illustrated in previous chapters, the majority of fishery stakeholders strongly agree that it is time to close the frontier. Alternatives for limiting access to recreational fisheries need to be considered as well.
- **Resolve the Individual Fishing Quota Moratorium**: The MSFCMA, as currently written, imposes a moratorium on the submission or approval of new individual fishing quota programs from January 4, 1995, until October 1, 2000. Because of the potential of such programs to assist in reducing capacity in the fisheries, an immediate decision whether to repeal this moratorium and return individual fishing quotas as a tool to the regional fishery management councils is imperative.

- **Develop Clear Allocation Criteria:** As fishery resources continue to be divided among an increasing number of participants, there is a critical need to develop clear allocation criteria to help fishery managers decide who gets the fish. This is particularly true for allocation between commercial and recreational fishermen. Guidance for such criteria could be provided in the MSFCMA or be developed by the regional fishery management councils.

- **Clarify the Respective Roles of the Regional Fishery Management Councils, Their Advisers, the National Marine Fisheries Service, the Secretary of Commerce, and Congress:** The public should expect stewardship from fishery managers and compliance from user groups. Participants in the fishery management system should expect reasoned discussion, respectful behavior, and accountability in public discussions. To address the issue of accountability, policymakers need to clarify the various jurisdictions and authorities in management—to be specific about who has control over which areas and why. The role, and use, of scientific and statistical committees within each council region should be further defined. Separating conservation decisions from allocations is becoming increasingly important as pressures on the management system have increased. This is not a new question, but it is one that has never been resolved.

- **Diversify Participation in the Regional Fishery Management Council System:** Lawmakers and decisionmakers need to decide how to better involve nontraditional interests in the regional fishery management council process, such as environmental groups, small-scale recreational fishermen, academics, and seafood consumers. Options to consider include changing the guidance to state governors about types of interests to represent, removing appointments from governors' offices, and designating council seats to different interests.

- **Consider User Fees as a Means of Cost Recovery or to Provide Returns to the Public:** At present, the MSFCMA prohibits the collection of fees over and above administrative costs, except for individual fishing quota fisheries, where a special fee limited to no more than 3.5 percent of the value of the quota is to be imposed. Allowing for the collection of user fees to assist in the costs of fishery management, enforcement, and/or research would present a real form of assistance to fishery managers and scientists in a time of scarce resources. User fees as a means of providing returns to the public could also be considered.

- **Define the Role of Industry in Data Collection and Research:** Industry can be an effective collaborator in research and data collection. For example, Nova Scotia's Fishermen and Scientists Research Society operates to foster this collaboration. Fishery managers and scientists should define areas where they can best collaborate with industry and develop a framework by which to structure and fund this collaboration.
- **Strengthen the Biological Scientific Basis for Fishery Management:** The biological science that informs fishery management decisions needs to be strengthened in both quantity and quality. Maximum sustainable yield regained central importance in fish stock assessments and management as a result of the 1996 MSFCMA reauthorization, but using it as a management benchmark is problematic in fish stocks about which little is known. In addition, recent MSFCMA provisions and implementing guidelines have begun directing U.S. marine fishery management toward an ecosystem-based approach. For these reasons, managers must be prepared to obtain and consider much more biological information from sources that may be nontraditional, such as industry at-sea observations. Fishery managers and scientists must also be prepared to deal with evaluating and integrating ecosystem-based management tools, such as marine protected areas.
- **Strengthen the Social Scientific Basis for Fishery Management:** Deciding who gets the fish dominates fishery management council meetings. It is crucial to develop a framework in which to analyze the social and economic issues that accompany these decisions. This will require the funding and strengthening of social science data collection and social science staff within the councils and the National Marine Fisheries Service. The existing ban on collecting proprietary economic information and the requirements to obtain approval for all data collected from the public are major hindrances to improving the social scientific basis of fishery management. When systematic social science analysis is absent, councils are forced to guess about the social and economic impacts of their decisions or rely on those members of the public who present testimony to the council. Neither approach provides a systematic assessment.
- **Develop an Evaluation Framework:** Fishery managers should develop a framework to routinely and systematically evaluate the performance of fishery management plans. If such a framework is

developed, it will need to have the flexibility for adaptation. Learning about fishery management and the effectiveness of management measures will require resources and time, and also some changes in the MSFCMA to provide the authority.

- **Strengthen Education and Training:** Council members need training to do their jobs and participants need information and education that will help them engage effectively in the fishery management process. The National Marine Fisheries Service and the Congress should develop a plan to foster better decisionmaking through professional development, education, and training for agency staff, council members, and fishery managers. Mandatory orientation and training of new council members should also be considered as a way to foster better decisionmaking and informed leadership.
- **Bring Fishery Management Incentives into Line with Long-Term Goals for Fisheries:** A systematic review of the incentives and disincentives in the current fishery management system should be undertaken to identify disincentives and unnecessary subsidies. Where possible, alternatives for incentive-based solutions to fishery management problems should be identified. Experimentation within council regions should be encouraged.

Conclusion

The only way we can manage is to make the fishing industry an integral part of the system. But we have a long way to go both technically and institutionally to accomplish this.

Ray Hilborn, University of Washington

We all believe in the goals of having enough fish for tomorrow, and having enough fishermen and processors around for tomorrow. If we don't learn to work together, none of us will be in business 10 years from now.

Rod Moore, West Coast Seafood Processors Association

Some fishery stakeholders with considerable experience are of the opinion that for the next few years we should focus on implementing the 1996 amendments, rather than taking on new tasks. The focus should include clarifying our overall policy and the definitions that we use. Others just as credible argue that the MSFCMA should be reexamined to see if it is accomplishing intended objectives. They view the act's existing provisions as impractical and believe that they cannot be accomplished, arguing that the "one-size-fits-all" provisions are not amenable to each and every region because they result from congressional micromanagement in the interest of specific regional interest groups.

The issues we propose for policy discussion and choice offer a little of both: some creative examination of fundamental principles as well as some hard-nosed evaluation of fishery management and policy performance. Even if the MSFCMA cannot be all things to all people, it will take all the voices from our fisheries to shape it in a way that provides for the transition into the future of U.S. fishery management.

In Their Own Words*

*These statements reflect the personal views of interview participants and do not necessarily represent the positions of the institutions or organizations with which they are associated.

Jesse Chappell, President, National Aquaculture Association

Federal support and funding for research, technical and management expertise, marketing and promotion are necessary for the continued growth of U.S. aquaculture. Areas of concern include jurisdictional regulation of aquaculture by the Departments of Commerce, Agriculture, and Interior as well as other independent agencies; development of a national aquatic animal health management program that well serves commercial aquaculture interests; fish-eating bird control; aquatic animal welfare; environmental stewardship; quality assurance; availability of aquatic animal therapeutants, chemicals, and vaccines; a more equitable treatment of U.S. aquacultured product versus that of other countries—a level playing field to market aquaculture products, and recognition of aquaculture as agriculture in all federal and state arenas.

Dorothy Childers, Executive Director, Alaska Marine Conservation Council

The big choices facing fishery management are how to deal with overcapitalization and how to build an ecosystem-based management system. In doing so, we must accept that there are limits to science and proceed under the precautionary principle. Since getting to an ecosystem-based system is a process, each decision along the way will either get us closer or keep us away from the goal. We have to be mindful of the steps we make along the way because the little steps add up to big steps.

Francis Christy, Senior Research Officer, IMARIBA

We need to bite the bullet and make a choice to manage fisheries under satisfactory property rights rather than fiddling around and avoiding the issue. Policy should facilitate moving toward these rights. There are social problems, but they are magnified beyond their importance. You can't make an omelet without breaking eggs—it can't be avoided. There should be something to alleviate the hardship, however. I do have some sympathy for the small-scale fishermen out there facing nature.

Doug Hopkins, Senior Attorney,
Environmental Defense Fund

A trend now is to put greater emphasis on the social impacts of fishery management. I think the intent is to have government be better at preventing the disruption of fishing communities and achieving greater equity. There is a mistaken notion that we can achieve all objectives at once and that government is going to do a good job with the social engineering of communities. This is the wrong direction to be moving in. There is no role for government in achieving certain social engineering objectives. Government shouldn't try to micromanage and freeze in time communities indefinitely. Communities evolve. Managers should instead put greater emphasis on assembling, analyzing, and presenting the economic implications—long-term versus short-term—of management decisions to com-munities.

Paul Howard, Executive Director,
New England Fishery Management Council

I don't think that the new law is flexible enough to account for the unique and individual aspects of all the different fisheries in the United States. The Sustainable Fisheries Act is a "sweeping" set of laws and requirements for the regional fishery management councils, some of which are totally unrealistic given our current knowledge base. This has led to ineffective fisheries management as well as rash management decisions that have been made with very little information or very little time to consider all the different aspects of individual fisheries. The deadlines and requirements mandated by the Sustainable Fisheries Act could ultimately lead to a step in the wrong direction.

Roger McManus, Vice Chair for Ocean Policy, Center for Marine Conservation

We need to look as a society at what we're going to do for the fishing industry. Here's a group of people facing a major life change. The industry won't go back to what it was. There is no more open frontier. The fishery will be increasingly regulated and conducted as a business. Economics will win out. Those who succeed will accumulate wealth and those who fail will not, just like regular life. Should we let them sink or swim? Should we try to freeze time? We shouldn't force anyone to become a museum specimen. We should treat people with respect. If we want to make the social choice to preserve a way of life or way of doing business, that's fine. But it should be a conscious choice about what decision is being made, not confused with something else. I would like to see the social objectives and choices made in consideration of the people affected.

J. Walter Milon, Professor of Environmental and Resource Economics, University of Florida

Reauthorization should look at the appropriate role of rights-based management. There must be something done to reduce the reluctance to embrace rights-based approaches. Congress said "STOP" in 1996 and it was a big mistake. Some experiments were occurring that could have provided some useful information in this area. There needs to be a lot more learning by doing. State and federal, recreational, and other interests should be more sensitive to broader-based ecosystem management. We need to examine the adequacy of resources to do fisheries research and management, and we should consider the establishment of an external review board of some kind— a neutral authority.

Michael K. Orbach, Director, Marine Laboratory, Nicholas School of the Environment, Duke University

We have not taken the case for fishery management and conservation outside the "club" of fishers and fish scientists. We need to bring the general public into these issues. For example, we are unlikely to make significant progress on habitat issues without involving a wider public. Many of the issues reach inland, and people there must understand the implications of their actions. And they must support the solutions and the government resources to get those solutions.

Carl Safina, Vice President for
Marine Conservation, National Audubon Society

The problems we've had and the various opportunity costs we've incurred, to say nothing of the ecological costs, all point in the same direction: after an era of development, which overshot and brought about a lot of serious depletions, we need to enter an era of rebuilding. However, this era—if it will be an era—has already begun, because we have some really major success stories, like striped bass and summer flounder for example. If nothing else, they should make us feel really hopeful that these things are controllable and respond to management and decisionmaking in the right direction and point the way for what kinds of decisions and mental postures to the questions are the right ones.

Michael L. Weber, Freelance Writer and
Researcher on Marine Conservation Issues

The major policy choice relates to reducing fishing capacity— if that is not solved then all other bets are off. Second, we need to address closing of the frontier. History shows that we are heading in that direction. We'll need to manage the inevitable transition humanely. It would be helpful to encourage experimentation in reauthorization rather than getting into the level of micromanagement that occurred the last time around. That's been a pattern over the years, but the suspension of ITQs was particularly egregious and unwarranted, particularly since it was largely a controversy from one or two regions of the country. It completely contradicted the notion of regional management. We have to start making choices instead of operating with the paradigm that we can maximize an unlimited number of objectives. That is logically impossible.

APPENDIX A

PROGRAM PARTICIPANTS

Interview Sources

Government

Robin Alden, former commissioner, Maine Department of Marine Resources

Neal Coenen, marine regional supervisor, Oregon Department of Fish and Wildlife

John H. Dunnigan, executive director, Atlantic States Marine Fisheries Commission

Peter Fricke, social anthropologist, Office of Sustainable Fisheries, National Marine Fisheries Service

Graciela García-Moliner, special assistant to the executive director, Caribbean Fishery Management Council

Bill Gordon, retired rancher and conservationist, former assistant administrator for Fisheries, National Marine Fisheries Service

John Green, former commissioner, Texas Parks and Wildlife Department, former chair, Gulf of Mexico Fishery Management Council

Jim Harp, Quinault Indian Nation, member, Pacific Fishery Management Council

Margaret F. Hayes, assistant general counsel for Fisheries, National Oceanic and Atmospheric Administration

Paul Howard, executive director, New England Fishery Management Council

Rick Lauber, chair, North Pacific Fishery Management Council

Vince O'Shea, captain, U.S. Coast Guard, 17th District, Juneau, Alaska

Sam Pooley, industry economist, Southwest Fisheries Science Center, Honolulu Laboratory, National Marine Fisheries Service

Andrew A. Rosenberg, deputy assistant administrator for Fisheries, National Marine Fisheries Service

Dick Schaefer, chief, Intergovernmental and Recreational Fisheries, National Marine Fisheries Service

Rolland A. Schmitten, former assistant administrator for Fisheries, National Marine Fisheries Service

Bob Schoning, resource consultant, former director, National Marine Fisheries Service

Michael Sissenwine, director, Northeast Fisheries Science Center, National Marine Fisheries Service

Gary Stauffer, director, Resource Assessment, Conservation and Engineering Division, Alaska Fisheries Science Center, National Marine Fisheries Service, member, Pacific Fishery Management Council Science and Statistical Committee

Environment

Eugene C. Bricklemyer Jr., president, Aquatic Resources Conservation Group

Paul Brouha, associate deputy chief of the National Forest System, U.S. Forest Service, former executive director, American Fisheries Society

Scott Burns, director, Marine Conservation Program, World Wildlife Fund

William J. Chandler, vice president for Conservation Policy, National Parks Conservation Association

Dorothy Childers, executive director, Alaska Marine Conservation Council

Jerry E. Clark, director of Fisheries National Fish and Wildlife Foundation

Sonja Fordham, fisheries project manager, Center for Marine Conservation

Rod Fujita, marine ecologist, Environmental Defense Fund

Burr Heneman, marine policy consultant

Ken Hinman, president, National Coalition for Marine Conservation

Doug Hopkins, senior attorney, Environmental Defense Fund

Gerald Leape, legislative director, Ocean Issues, Greenpeace

Roger E. McManus, vice chair for Ocean Policy, Center for Marine Conservation

Carl Safina, vice president for Marine Conservation, National Audubon Society

Cynthia Sarthou, director, Gulf Restoration Network

Peter Shelley, vice president, Conservation Law Foundation

Lisa Speer, co-director, Ocean Protection Initiative, Natural Resources Defense Council

Michael L. Weber, freelance writer and researcher on marine conservation issues

Industry

Linda Behnken, executive director, Alaska Longline Fishermen's Association, member, North Pacific Fishery Management Council

Charlie Bergmann, Lund's Fisheries, member, Mid-Atlantic Fishery Management Council, member, New Jersey Marine Fishery Council

Ray Bogan, legal counsel, United Boatmen of New Jersey and New York

Jesse Chappell, president, National Aquaculture Association

Jim Cook, chair, Western Pacific Fishery Management Council, captain, F/V *Katy Mary*, Honolulu, Hawaii

Don DeMaria, captain, F/V *Misteriosa*, Summerland Key, Florida

Barry Fisher, president, Midwater Trawlers Cooperative, managing owner, F/V *Excalibur*, Newport, Oregon

David Fraser, captain, F/V *Muir Milach,* Port Townsend, Washington

Jim Gilmore, director of public affairs, At-sea Processors Association

Thomas Hill, chair, New England Fishery Management Council

Jim Kendall, executive director, New Bedford Seafood Coalition

Dave Krusa, captain, F/V *Restless,* Montauk, New York

John A. Mehos, former president, Texas Shrimp Association

Rod Moore, executive director, West Coast Seafood Processors Association

Mike Nussman, vice president, American Sportfishing Association

Gil Radonski, retired president and chief executive officer, Sport Fishing Institute

Angela Sanfilippo, president, Gloucester Fishermen's Wives Association

Bob Storrs, vice president, Unalaska Native Fisherman's Association, member, Alaska Marine Conservation Council, captain, F/V *Flying Oosik,* Unalaska, Alaska

Roger Thomas, president, Golden Gate Fishermen's Association, captain, F/V *Salty Lady,* Sausalito, California

Lee J. Weddig, former executive vice president, National Fisheries Institute

Richard Young, captain, F/V *City of Eureka,* Crescent City, California

Academia

Dayton L. Alverson, chairman of the board, Natural Resources Consultants, Inc.

Lee G. Anderson, professor, College of Marine Studies, University of Delaware

William T. Burke, professor of law, University of Washington

Francis Christy, senior research officer, IMARIBA

Jim Crutchfield, professor emeritus of economics and marine affairs, University of Washington

Christopher M. Dewees, marine fisheries specialist and Sea Grant Extension Program coordinator, Department of Wildlife, Fish, and Conservation Biology, University of California, Davis

Robert B. Ditton, professor of wildlife and fisheries sciences, Texas A&M University

Michael J. Fogarty, associate professor, Chesapeake Biological Laboratory, University of Maryland Center for Environmental Science

Timothy Hennessey, professor of political science and professor of marine affairs, University of Rhode Island

Ray Hilborn, professor, School of Fisheries, University of Washington

James E. Kirkley, associate professor of marine science, College of William and Mary, Virginia Institute of Marine Science

J. Walter Milon, professor of environmental and resource economics, University of Florida

Michael K. Orbach, professor of marine affairs and policy and director, Marine Laboratory, Nicholas School of the Environment, Duke University

Terrance J. Quinn II, professor, Juneau Center, School of Fisheries and Ocean Sciences, University of Alaska

Alison Rieser, director, Marine Law Institute, School of Law, University of Maine

Brian J. Rothschild, director, Center for Marine Science and Technology, University of Massachusetts, Dartmouth

David Sampson, associate professor of fisheries, Oregon State University

Courtland L. Smith, professor of anthropology, Oregon State University

James Wilson, professor of resource economics and associate director, School of Marine Sciences, University of Maine

Reviewers
Government

Gene Buck, senior analyst, Congressional Research Service

Randy Fisher, executive director, Pacific States Marine Fisheries Commission

Pamela M. Mace, fisheries scientist, Office of Science and Technology, National Marine Fisheries Service

Ben Muse, economist, Alaska Commercial Fisheries Entry Commission

Robert Palmer, chief, Bureau of Marine Fisheries Management, Division of Marine Fisheries, Florida Fish and Wildlife Conservation Commission

Clarence Pautzke, executive director, North Pacific Fishery Management Council

Steven Pennoyer, regional administrator, Alaska Regional Office, National Marine Fisheries Service

William S. "Corky" Perret, director, Office of Marine Fisheries, Mississippi Department of Marine Resources

Robert White, former administrator, National Oceanic and Atmospheric Administration

Environment

Robert Eaton, director, Pacific Marine Conservation Council

Paul Engelmeyer, NW policy analyst, Living Oceans Program, National Audubon Society

Rebecca J. Goldburg, scientist, Environmental Defense Fund

Anne Platt McGinn, senior researcher, Worldwatch Institute

Vikki Spruill, executive director, SeaWeb

Patricia Tummons, editor, Environment Hawai'i

Industry

Richard Gutting, executive director, National Fisheries Institute

Peter Leipzig, executive director, Fishermen's Marketing Association

Mark Lundsten, captain, F/V *Masonic*, Seattle, Washington

Ellen Peel, director, The Billfish Foundation

Brooks Takenaka, manager, United Fishing Agency, Ltd., Honolulu, Hawaii

Academia

Felicia Coleman, research faculty, Department of Biological Science, Florida State University

David L. Fluharty, research associate professor, School of Marine Affairs, University of Washington

Caroline Pomeroy, assistant research scientist, Institute of Marine Sciences, University of California, Santa Cruz

LaVerne Ragster, vice president, Research and Public Service, University of the Virgin Islands

Ralph E. Townsend, chair, Department of Economics, University of Maine

ANNOTATED BIBLIOGRAPHY OF SELECTED LITERATURE ON U.S. MARINE FISHERY MANAGEMENT

Fishery Status and Management History

Atlantic States Marine Fisheries Commission, Gulf States Marine Fisheries Commission and Pacific Marine Fisheries Commission. 1977. *A Report to the Congress: Eastland Fisheries Survey, May 1977.* Portland, OR: Pacific Marine Fisheries Commission.

> This report, commissioned by Congress, is the result of an 18-month study by the three commissions to assess the problems and needs of U.S. fisheries. It provides policy recommendations developed by national consensus at the National Conference held November 29–December 2, 1976, related to fishery conservation and management, utilization and development, recreation, and the special needs of island territories. It also provides regional recommendations not addressed by the conference, supported by excerpts from regional reports. The report concludes with a summary of introductory remarks made at the conference.

Conrad, J. M. 1987. The Magnuson Fisheries Conservation and Management Act: An Economic Assessment of the First 10 Years. *Marine Fisheries Review* 49(3):3–12.

> This paper reviews the background and enactment of the FCMA and management policies related to the act. It discusses the theory of open access and assesses the act's impacts on the economic performance of the fishing industry from 1977 to 1985. The assessment indicates that, although landings and net revenues increased significantly for the U.S. fishing industry from 1977 to 1983, excess capacity is leading the industry toward an open access equilibrium and the ultimate dissipation of all rents. The paper concludes with recommendations to improve the situation.

Cunningham, S. 1981. The Evolution of the Objectives of Fisheries Management During the 1970s. *Ocean Management* 6:251–78.

> This paper describes the major biological, economic, and political objectives of fisheries. It begins with a description of the bioeconomic model and a discussion of why marine fishery management is essential. Next, it describes and critiques each of various fishery management objectives considered in the 1970s, noting the advantages and problems of each. These include maximum sustainable yield, optimum sustainable yield, and static and dynamic maximum economic yield. The author briefly discusses adaptive control of fisheries and concludes with a look at the political objectives of fishery management, including income distribution, employment, balance of payments, and maximization of food production.

Dewar, M. E. 1983. *Industry Trouble: The Federal Government and the New England Fisheries.* Philadelphia, PA: Temple University Press.

This book reviews and analyzes the role of the federal government in New England fisheries post–World War II. The author examines both the rationale behind federal involvement and the effects of this involvement. She begins with a description of the structure and problems of the offshore groundfish industry. She then describes the industry campaign for government aid that led to the many federal programs of the fifties and sixties and discusses the shortcomings of the subsequent intervention that resulted from misconceptions about the industry and incorrect problem identification. Finally, she discusses the failure of government intervention in the groundfish fishery more broadly as a case study of government efforts to rescue distressed industries through pricing policy, tariffs, subsidies, and regulation.

Hargis, W. J., Jr., R. J. Baker, F. Bemiss, J. C. Cato, J. P. Harville, A. W. Haynie, H. Lyman, J. A. Mehos, J. G. Peterson, C. L. Smith, and W. E. Towell. 1986. *NOAA Fishery Management Study*. Washington, DC: National Oceanic and Atmospheric Administration, U.S. Department of Commerce.

This report was solicited by the National Oceanic and Atmospheric Administration to define the goals of marine fishery management, to consider alternative institutional arrangements and management strategies to meet these goals, to identify and prioritize activities to implement management strategies, and to specify areas of public and private responsibility and accountability. The authors recommend major conceptual and operational changes for marine fisheries—including a national-level determination of conservation decisions and a regional-level determination of allocation decisions, subject to secretarial review. Other changes relate to the number of councils within the regional council system and the process of nominations for council membership, interjurisdictional management, the need for more and better data, funding priorities, user fees and licenses, limited entry, full domestic utilization, and habitat.

Larkin, P. A. 1984. The Problem with George, or The Role of Development in Fisheries Management. *Marine Recreational Fisheries* 9:111–15.

This insightful article tells the story of a new fishery manager's coming to terms with the responsibility of simultaneously protecting, regulating, and developing recreational fisheries. The story focuses on the consequences of unchecked development and asks the familiar question, Is there such a thing as overdevelopment of a fishery resource?

National Marine Fisheries Service. 1976. *A Marine Fisheries Program for the Nation*. Washington, DC: National Oceanic and Atmospheric Administration, U.S. Department of Commerce.

This report discusses the status and future outlook of U.S. marine fisheries in 1976 and outlines the goals of the Commerce Department at that time: "[T]o conserve and manage resources; conserve, restore, and enhance fish habitats; develop and maintain a healthy commercial fishing industry; strengthen the contribution of marine resources to recreation; encourage the development of public and private aquaculture; and assure the safety, quality, and identity of seafoods." The report provides recommendations in response to these goals, many of which were incorporated into the FCMA. Each recommendation references corresponding sections of the act. The National Plan for Marine Fisheries is included as an appendix.

Pautzke, C. G. 1992. *The Magnuson Act after Seventeen Years: Is It Working?* Anchorage, AK: North Pacific Fishery Management Council.

This paper, developed in anticipation of the 1993 reauthorization of the MFCMA, favorably discusses the effectiveness of the MFCMA and the regional council system in managing U.S. marine fisheries. The condition of national fisheries pre-1976, as defined by comptroller general reports, is compared to that post-1976 and the MFCMA. The author highlights the implementation of the act with remarks recorded at the 1976 National Conference for Regional Councils and discusses issues of authority between the councils and NMFS. Achievements of the North Pacific Fishery Management Council are outlined from 1976 to 1992, and the paper concludes with a discussion of the strengths and weaknesses of the regional council system.

Rettig, R. B. 1987. The Magnuson Fishery Conservation and Management Act Revisited. *Marine Policy Reports* 10(1):6–11.

This report reviews the post-war period leading to the implementation of the FCMA and discusses U.S. fisheries under the act in terms of controlling foreign fishing effort and establishing a new domestic fishery management regime. The complexity of state and federal roles and responsibilities is discussed and the "new" interest in limited entry briefly addressed. Controversial studies/reports related to amending the act are highlighted, and the report concludes with a brief discussion of the concept of adaptive management.

Wise, J. P. 1991. *Federal Conservation and Management of Marine Fisheries in the United States.* Washington, DC: Center for Marine Conservation.

This publication discusses the MFCMA's effects on U.S. marine fishery conservation and management. Important provisions of the act are described in detail, as are the 32 fishery management plans active at the time of publication. Major international fisheries treaties and agreements are reviewed. The author highlights conclusions and recommendations from major U.S. marine fishery studies since 1977 and depicts the fragmentation of federal fishery management programs and budgets. He concludes with seven recommendations related to furthering the application of conservation principles in the act. Appendices include information on fisheries population biology and management, a chronology of policy affecting U.S. marine fisheries, miscellaneous statistical tables, a list of applicable nongovernmental organizations, the Dingell-Johnson/Wallop-Breaux Federal Funding for Marine and Anadromous Fisheries Projects through FY'89, and diagrams illustrating fishery management plan formulation and implementation.

Assessing Fishery Productivity

Anderson, L. G. 1980. Necessary Components of Economic Surplus in Fisheries Economics. *Canadian Journal of Fisheries and Aquatic Sciences* 37(5):858–70.

This paper introduces and discusses four main types of economic surpluses resulting from the exploitation of a fishery: (1) industry profit, (2) normal factor rents, (3) consumer surplus, and (4) worker satisfaction bonus. The fourth component is discussed in detail, as are the implications of this component for fishery management and policy. The author develops a detailed model to allow for the joint consideration of these components and to show how they vary at different levels of effort.

Barber, W. E. 1988. Maximum Sustainable Yield Lives On. *North American Journal of Fisheries Management* 8(2):153–57.

This study examines the use of maximum sustainable yield in 142 different scientific papers from 1977 to 1985. It reports that the concept has been used primarily in the estimation of long-term yield, the evaluation of stock condition, and policy analysis. The author suggests two main reasons for the continued use of maximum sustainable yield in fisheries, despite its weaknesses. The first relates to the simplicity of the concept and the second to scientific and technical restrictions. He stresses the importance of developing a protocol for using maximum sustainable yield in the decisionmaking process.

Ecosystem Principles Advisory Panel. 1999. *Ecosystem-Based Fishery Management: A Report to Congress as Mandated by the Sustainable Fisheries Act Amendments to the Magnuson-Stevens Fishery Conservation and Management Act, 1996.* Silver Spring, MD: National Marine Fisheries Service.

This report is the result of a 1996 congressional charge to the National Marine Fisheries Service to form a panel for the purpose of evaluating the current application of ecosystem-based fisheries management and developing recommendations for further implementation of ecosystem-based approaches within the context of the existing fishery management system. The report identifies basic principles, goals, and policies for ecosystem-based fishery management and provides recommendations related to developing and implementing fisheries ecosystem plans and the research required to support management.

Edwards, S. F. 1990. *An Economics Guide to Allocation of Fish Stocks between Commercial and Recreational Fisheries.* NOAA Technical Report NMFS 94, November 1990. Woods Hole, MA: National Marine Fisheries Service Northeast Fisheries Center.

This report offers the noneconomist a guide to the economic concepts, principles, and methods that are fundamental to understanding the economic implications of fishery allocations between competitive uses. The author defines economic value, presents the elements of both benefit-cost and input-output analyses, and compares the two methodologies in the context of fishery valuation. He then provides an example of an efficient catch allocation between the commercial and recreational sectors. Appendices include a glossary as well as more detailed considerations of input-output analysis and valuation of fishery resources in a multimarket framework.

Holden, M. 1994. The Conservation Policy: Fisheries Conservation or Fisheries Management? or Biology v. Economics. In *The Common Fisheries Policy: Origin, Evaluation and Future,* 169–85. Cambridge, MA: Fishing News Books.

In an attempt to reconcile some common misunderstandings and misconceptions, this paper describes the basic biological principles—life history, growth and mortality, spawning stock biomass, and year-class size—underlying fishery conservation and reviews the fisheries model on which management is based—yield per recruit models, total allowable catches, and multispecies models. Both the definitions and objectives of fishery "conservation" and "management" are distinguished, as are the implications of biological and economic goals.

Larkin, P. A. 1977. An Epitaph for the Concept of Maximum Sustained Yield. *Transactions of the American Fisheries Society* 106(1):1–11.

This article briefly describes the evolution of the concept of maximum sustainable yield—from development to global acceptance. It outlines the biological and economic implications of the concept and describes why maximum sustainable yield is impossible to achieve on a sustained basis, for single- or multi-species populations. The author introduces the concept of optimum sustained yield and describes the difficulties encountered in trying to define this term or use it as an operational basis for making decisions. He concludes with potential options for the future and a witty epitaph.

Rosenberg, A. A., M. J. Fogarty, M. P. Sissenwine, J. R. Beddington, and J.G. Shepherd. 1993. Achieving Sustainable Use of Renewable Resources. *Science* 262(5135):828–29.

This short essay was written in response to an earlier article in *Science* by Ludwig, Hilborn, and Walters. The authors argue that we can indeed achieve sustained use of renewable resources and that there exists historical evidence of resource sustainability. They review the theoretical and empirical basis for sustainable yield, discuss obstacles to achieving sustainability, and suggest remedies for these obstacles. Finally, they conclude with an examination of evolving trends in scientific advice.

Sherman, K., and L. M. Alexander. 1986. *Variability and Management of Large Marine Ecosystems*. Boulder, CO: Westview Press, Inc.

This book is a compilation of scientific articles related to the variability and management of large marine ecosystems. The first section describes the impact of human and environmental perturbations on the productivity of renewable resources in these ecosystems. The second section discusses strategies for measuring and forecasting the natural variability of large marine ecosystems against a background of human-induced pollution and overexploitation. And the final section reviews the institutional framework for managing large marine ecosystems.

Smith, C. L. 1980. Management Provoked Conflict in Fisheries. *Environmental Management* 4(1):7–11.

This paper describes how the application of fishery management concepts involves social goals that lead to frustration and conflict in fisheries. The author points out the conservation preferences inherent in maximum sustainable yield and shows how acceptable amounts of variability are related to social goals. He discusses the expectations associated with capacity and describes how these result in conflict. He also stresses the need for a link between publicly identified objectives and goals implied in management concepts.

Toman, M. A. 1994. Economics and "Sustainability": Balancing Trade-Offs and Imperatives. *Land Economics* 70(4):399–413.

This article discusses the concept of sustainability from an economic, ecological, and social perspective. The author provides terms that may be useful for the debate between disciplines and identifies two issues as central themes in any conception of sustainability: (1) intergenerational fairness, and (2) resource substitutability. Alternative views on these issues are examined to show how they lead to different conceptions of sustainability. Finally, these alternative conceptions are related to each other through a simple conceptual framework of resource management based on the idea of "safe minimum standard." This model allows the consideration of individual trade-offs and social imperatives.

Owning Fishery Resources

Christy, F. T. 1996. The Death Rattle of Open Access and the Advent of Property Rights Regimes in Fisheries. *Marine Resource Economics* 11:287–304.

This paper proposes that fishery research and administration associated with resolving problems of open access are wasteful and that attention should, instead, be turned toward a new paradigm of property rights regimes. The author begins with a discussion of present paradigms based on the prevailing condition of open access and the costs to researchers and stakeholders of maintaining these paradigms. He then discusses property rights regimes as the paradigms of the future and identifies possible obstacles and imperfections. He concludes with suggestions for future research.

Feeny, D., F. Berkes, B. J. McCay, and J. M. Acheson. 1990. The Tragedy of the Commons: Twenty-two Years Later. *Human Ecology* 18(1):1–19.

This article critiques Hardin's famous thesis about the commons based on common-property resource management evidence collected after the presentation of Hardin's argument in 1968. The authors begin with the identification and discussion of four categories of property-rights regimes within which common-property resources are held: (1) open access, (2) private property, (3) communal property, and (4) state property. They then provide examples of management under each regime focusing first on exclusion, then on regulation of use and users. They conclude that private, communal, and state property regimes are all potentially viable resource management options and that Hardin's argument overlooked the important role of institutional arrangements and cultural factors.

Grafton, R. Q., D. Squires, and J. E. Kirkley. 1996. Private Property Rights and Crises in World Fisheries: Turning the Tide? *Contemporary Economic Policy* 14:90–99.

This paper explores the potential of privatization as a means to address the world fisheries crises. Several types of rights-based management are introduced, including territorial use rights, individual transferable quotas, and cooperative/community fishery management. The authors assess the performance of individual transferable quotas with respect to monitoring and enforcement, allocation rights, economic benefits, adjustments in the fishery, and resource rents. They discuss fluctuating fish stocks, straddling stocks and high seas fisheries, and the endemic poverty of many artisanal fisheries as issues unresolved by individual transferable quotas. They conclude that individual transferable quotas offer a significant improvement over many current fisheries practices but should not be considered applicable to all fisheries or as a panacea to all problems.

Hardin, G. 1968. The Tragedy of the Commons. *Science* 162(5364):1243–48.

This article discusses the controversial problem of unchecked population growth in a finite world. The author predicts that the
inevitable result of such growth is overexploitation and degradation.
He introduces the term "tragedy of the commons" to describe competition in an unmanaged common, where each participant acts to
maximize individual personal gain to the detriment of the other participants who share the negative effects of the single individual's
actions. He states that "freedom in a commons brings ruin to all"
and supports this statement with an example of a grazing regime,
where cattle are continuously added to a pasture by herdsmen
despite spatial limitations. He also extends this theory to the oceans
and our national parks system. Finally, he discusses issues of con-
science and mutual coercion, arguing for abandonment of the commons and a shift toward some sort of property rights.

Kaufmann, B., and G. Geen. 1997. Cost-Recovery as a Fisheries Management
Tool. *Marine Resource Economics* 12:57–66.

This paper contends that the current failure of fishery management
results from the use of public institutions to deliver fishery management services. The authors argue that the incentives for efficiency
inherent in a properly designed cost-recovery policy will result in
more effective institutional and operational arrangements. They discuss problems related to the implementation of such a policy, including opposition from fisheries managers, researchers, and industry.
They describe some of the effects of cost-recovery in Australia, provide a brief overview and comparison of existing cost-recovery
regimes in Australia and Canada, and highlight important lessons
from these experiences related to the design of the policy and the role
of finance/treasury departments in ensuring proper implementation
of the program.

Mace, P. M. 1993. Will Private Owners Practice Prudent Resource Management? *Fisheries* 18(9):29–31.

This article cautions against common blanket assertions about the
theoretical, biological, and economic benefits of property rights systems. The author argues that ownership alone may be an insufficient
mechanism to guard against overexploitation and that government
or other "publicly accountable" control is necessary. She also cautions about the difficulty of making fundamental program (or property rights definition) changes. She recommends that government
remain involved in establishing, monitoring, and enforcing fishery
catch limits and that program design is transitionary and changes
incremental. Her points are illustrated with empirical examples
from New Zealand fisheries.

McCay, B. 1995. Common and Private Concerns. *Advances in Human Ecology*
4:89–116.

This paper describes the "tragedy of the commons" perspective as both inaccurate and wrongfully leading to arguments for strong, centralized government or privatization. The author distinguishes between common pool and common property and discusses the importance of social and community factors in understanding the interactive dynamics of fisheries. She exemplifies the fisherman's problem of interdependence and open access with a discussion of the east coast summer flounder fishery. She proposes a more progressive way in which to order the study of "the commons" and suggests a broader and more complex range of alternatives than those offered by the "tragedy of the commons" perspective. These alternatives, labeled "comedies of the commons," include communal management, comanagement, and user participation. Finally, she concludes with some "private concerns" about the rapid leap from limited access to private property rights.

National Research Council. 1999. *The Community Development Quota Program in Alaska and Lessons for the Western Pacific.* Washington, DC: National Academy Press.

This report was produced in response to a 1996 congressional request to review the Alaska community development quota program. The report provides a comprehensive review of the program and its application to the rural western Alaska region, describes strengths and weaknesses of the program, outlines operational concerns, and discusses lessons that can be applied to the development of similar programs in other areas. It indicates that the western Alaska program offers great social and economic potential for the region but that such programs may not be as effective in other areas, such as the western Pacific, where local context and goals differ. Recommendations are provided.

National Research Council. 1999. *Sharing the Fish: Toward a National Policy on Individual Fishing Quotas.* Washington, DC: National Academy Press.

This report was produced in response to a 1996 congressional request to report on the social, economic, and biological effects of individual fishing quotas and other limited entry programs and to provide recommendations. It presents a comprehensive review of individual fishing quota systems, including their history, present use, benefits, and limitations. It examines the U.S. and foreign experience with these programs; looks at alternative conservation and management measures; addresses specific congressional questions about individual fishing quotas in relation to a set of "first principles," derived from the MSFCMA; and provides separate recommendations to Congress, the secretary of commerce, the National Marine Fisheries Service, the councils, the states, and others. It concludes that individual fishing quotas are valuable management tools and—although they should not be considered a panacea—they should be made available to the councils to use at their discretion, on a fishery-by-fishery basis.

Ostrom, E. 1990. *Governing the Commons: The Evolution of Institutions for Collective Action.* Cambridge: Cambridge University Press.

This book explores the institutional arrangements that lead to effective self-organization, self-governance, and resource management under a common-pool management regime. The author reviews the history of commons theory, provides an in-depth analysis of several long-standing and viable common-property regimes throughout the world, analyzes institutional change, and examines international cases of institutional failures. Finally, she concludes with a framework for analyzing self-organizing and self-governing common-property regimes.

Managing Fisheries

Cicin-Sain, B., and M. K. Orbach. 1986. Mutual Mysteries: Washington/Regional Interactions in the Implementation of Fisheries Management Policy. *Policy Studies Review* 6(2):348–57.

Based on 30 in-depth personal interviews with decisionmakers involved in producing the fishery management plan for California's northern anchovy fishery, the authors examine interorganizational differences between the national and regional levels responsible for implementing the MFCMA. Differences between organization location (formal status in the decisionmaking system) and socialization (professional background) at both levels have important impacts on implementation behavior and act to reinforce differences at each level in (1) the way laws are viewed, (2) what aspects of the legislation should be emphasized, (3) policy goals, (4) criteria for decisionmaking, and (5) perceptions of influence. This leads to confusion and misperception.

Hanna, S. S. 1997. The New Frontier of American Fisheries Governance. *Ecological Economics* 20:221–33.

This paper begins with a discussion of American resources as frontiers. The author compares the attributes of resource management under the frontier paradigm with that of the commons and describes behavioral differences between users in each model. She then describes the conditions required to develop the institutional capital necessary to achieve sustainable governance of U.S. fisheries. These involve viewing the fishery as an integrated ecosystem, identifying all shareholders, allocating decisionmaking power and responsibility to internalize control and vest all interests, developing incentives to promote long-term management, improving management skills, and utilizing adaptive management. She concludes with an assessment of the progress of U.S. fishery management in developing institutional capital.

Jentoft, S. 1989. Fisheries Co-management: Delegating Government Responsibility to Fishermen's Organizations. *Marine Policy* 13(2):137–54.

This article focuses on the division of responsibility between government and the fishing industry in fishery management and the ability of fishermen's cooperative organizations to perform regulatory functions. The author discusses the benefits, and potential problems, associated with a cooperative management approach and reviews international experiences with fishermen's organizations. He identifies critical variables necessary for the development of successful cooperative organizations, including legislation that enables the organizations to implement and enforce regulations, organizations small enough in scale to maintain manageability, relatively homogeneous socioeconomic membership, a tradition of cooperative and collective action, the organization's ability to establish trust with the fishermen, and an appropriate division of responsibility between government and industry.

Jentoft, S., and B. McCay. 1995. User Participation in Fisheries Management: Lessons Drawn from International Experiences. *Marine Policy* 19(3): 227–46.

This study, based on the findings of two international projects, examines the various ways in which user groups are incorporated in the fishery management process. The authors identify three institutional design alternatives that allow for user participation and summarize the various degrees of user participation in fishery management and the diversity of institutional design in 11 case studies. They develop lessons and conclusions from the case studies and address two questions of institutional design related to issues of representation and scale. Finally, they provide recommendations for future research.

King, L. R., and S. G. Olson. 1988. Coastal State Capacity for Marine Resources Management. *Coastal Management* 16:305–18.

As federal leadership and financial support for ocean programs diminish, coastal states have the potential to become major actors in ocean issues. This article identifies indicators of enhanced state capabilities in marine affairs and discusses the capacity—defined as a mature institutional structure, a high level of professional expertise, and the interest and commitment of political elites—of states to expand their involvement in coastal and marine policy issues. The authors find capability to vary greatly among coastal states, comparing the significant accomplishments of North Carolina with the failures of Texas. They discuss the roles of the Coastal Zone Management and Sea Grant College programs in expanding state institutional capabilities. Finally, they recommend further research in exploring and comparing the variable degree of effectiveness of state institutions and the establishment of networks to deal with coastal resource issues.

Lindblom, C. E. 1979. Still Muddling, Not Yet Through. *Public Administration Review* (November/December):517–26.

This discussion distinguishes incremental politics from incremental analysis, which is divided into categories of simple, disjointed, and strategic. The author argues that a strategic analytical approach is more productive than attempting to achieve complete scientific analyses of complex problems. He suggests that the incremental political approach is wrongly labeled as "slow-moving" and is often confused with problems resulting from the structure of veto powers and other political features characteristic of a market-oriented society. He describes partisan analysis as the most characteristic—and most productive—form of analytical input into politics. He argues that improved combinations of the three forms of incremental analysis, incremental politics, and partisan mutual adjustment and analysis can lead to more intelligent and democratically responsive policymaking.

Rose, A., B. Stevens, and G. Davis. 1989. Assessing Who Gains and Who Loses from Natural Resource Policy: Distributional Information and the Public Participation Process. *Resources Policy* 15(4):282–91.

This article describes how publicly administered resource allocation can be more effective and democratic when it is informed by data on the distribution of economic impacts. The authors specify three positive measures—the individual impact matrix, the community impact index, and the political articulation index—of the distribution of economic impacts, and they demonstrate an operational methodology that can be used to calculate these impacts. They use a case study of coal mining on public lands in the Monongahela National Forest to illustrate how even a net positive cost-benefit analysis can produce impacts that clash with citizen preferences. They conclude that successful public participation requires both the dissemination of information on individual policy impacts and the ability to predict public reaction.

Smith, M. E. 1990. Chaos in Fisheries Management. *Mast* 3(2):1–13.

This study briefly reviews the structure and composition of the regional fishery management councils and discusses how this participatory model complicates the councils' goals to sustain the stocks and optimize economic utilization. It explains how a participatory model that employs people with such widely diverse views of nature to manage fishery resources by consensus is likely to result in differing individual beliefs regarding what constitutes relevant data, how the data should be interpreted, and how appropriate management responses should be designed. It suggests that a better understanding of those factors dividing fishery managers and industry members is essential. Chaos theory and fisheries as both linear and nonlinear systems are discussed.

World Wildlife Fund. 1995. *Managing U.S. Marine Fisheries: Public Interest or Conflict of Interest?* Washington, DC: World Wildlife Fund.

This report details findings of a World Wildlife Fund investigation into conflicts of interest and the regional fishery management council system. Both the New England and the Gulf of Mexico fishery management councils are discussed independently with respect to status of stocks, species managed, and management history. Council members' actions on specific management issues—including roll call votes, motions, and even meeting comments—are illuminated and compared with information about their backgrounds and outside interests. The report concludes that "conflicts of interest plague the council process and contribute to well-documented shortcomings in the management of U.S. fisheries."

Creating Incentives

Anderson, L. G. 1989. Enforcement Issues in Selecting Fisheries Management Policy. *Marine Resource Economics* 6:261–77.

This article presents a framework useful in comparing fishery regulations and identifying enforcement issues important to the selection of fishery policy. The author discusses dockside versus at-sea monitoring, the ability of industry to comply with regulations, the risk of noncompliance, initial versus continued compliance, the degree to which benefits of compliance can be demonstrated, the potential for citizen cooperation in enforcement, the likelihood of regulations to promote rent-seeking behavior, and the effectiveness of enforcement with respect to management objectives. He also discusses the ease of regulations/enforcement with respect to government implementation—distinguishing between mistakes, bad practice, and cheating; communicating requirements; disguising noncompliance; detecting noncompliance such that it is admissible as evidence; detecting illegal activities under various conditions; and identifying benefit-based priorities.

Costanza, R. 1987. Social Traps and Environmental Policy: Why Do Problems Persist When There Are Technical Solutions Available? *BioScience* 37(6): 407–12.

This article describes the social trap phenomenon and examines ways to escape—or avoid altogether—these traps. The author characterizes these phenomena as situations in which short-term incentives are inconsistent with the long-term, best interest of the individual and society. He identifies the causes of social traps and describes four methods by which such traps can be avoided or escaped. These include (1) education, (2) insurance, (3) superordinate authority, and (4) converting the trap to a trade-off by charging the long-run costs associated with the activity to the responsible parties in the short run. He indicates that, while the latter method has been proven most effective experimentally, it has not been sufficiently applied in the environmental arena. He argues that converting traps to trade-offs, along with the other methods, can make the entire system more effective.

Fujita, R. M., D. Hopkins, and Z. Willey. 1996. Creating Incentives to Curb Overfishing. *Forum for Applied Research and Public Policy* 11(2):29–34.

These writers attribute the overexploitation of fishery resources to perverse economic incentives. They describe the disincentives for resource conservation created by open access fisheries, review the limitations of limited access regimes, and identify the strengths and weaknesses of individual transferable quota systems. They suggest cooperative- or co-management regimes as possible alternatives to open access and make several recommendations: managers should be required to consider the best scientific information in decision-making, a precautionary approach should always be prescribed, protection of ecological integrity should be a priority goal, conservative parameters should be required in fish population assessment, and total allowable catch should incorporate scientific uncertainty and natural variability.

Gordon, H. S. 1954. The Economic Theory of a Common-Property Resource: The Fishery. *Journal of Political Economy* 62:124–42.

The paper applies economic theory to the fishery and demonstrates that overfishing is a function of the economic organization of the industry. The author asserts that the role of fishers has been historically neglected in fisheries research, which has instead focused on biological production factors and theories. He introduces the concept of a net economic yield and shows that fishermen's behavior can be generalized in accordance with the standard economic theory of production. He argues that the inefficiency of fisheries production and the plight of fishermen result from the common property nature of fishery resources.

Hanna, S. S. 1998. Institutions for Marine Ecosystems: Economic Incentives and Fishery Management. *Ecological Applications* 8(1) Supplement:S170–74.

This article identifies some of the important economic dimensions of

resource management institutions that are embedded in both management structures and processes. The author argues that understanding these dimensions is a necessary part of expanding fishery
management to ecosystem scale because it provides managers the
opportunity to align economic incentives with ecosystem objectives.
She identifies several functions of institutions that properly coordinate the human use of ecosystems. These include management structures that promote the definition of multiple objectives and the
cost-effective coordination of organizational tasks, and management
processes that are legitimate, flexible, and promote socially appropriate time horizons.

Iudicello, S., M. Weber, and R. Wieland. 1999. *Fish, Markets and Fishermen:
The Economics of Overfishing.* Washington, DC: Island Press.

This book explains how biological, economic, and political components of fisheries link together to create incentives that shape fishing
behavior. The authors lead the reader to understand that overfishing
does not result from greed or weakness, but rather from perverse
incentives. Through theory and case studies, they show how economics drives fishing behavior, how subsidies hurt, how management tools work, and how economic tools can contribute to
solutions.

Larkin, P. A. 1988. The Future of Fisheries Management: Managing the Fisherman. *Fisheries* 13(1):3–9.

This paper identifies three components of fishery management:
habitat protection, enhancement of fish production, and regulation
of fishermen. The author asserts that fishermen should be regulated
in a manner that assists them in meeting their objectives, described
as self-satisfaction for sport fishermen, self- and family-support for
artisanal fishermen, and monetary for commercial fishermen. He
indicates that these socioeconomic objectives require a better understanding of human behavior and the marketplace.

Milazzo, M. 1998. *Subsidies in World Fisheries: A Reexamination.* World Bank
Technical Paper/Fisheries Series 406. Washington, DC: The World Bank.

This report describes subsidies as an important factor undermining
the sustainable use of fishery resources. It examines the implications
and impacts of fishing subsidies, focusing on impacts on the
resource, as opposed to trade. It reviews a wide range of indirect and
implicit assistance programs for the world's fishing fleets and provides estimates of the national and global impact of these programs.
Both budgeted and unbudgeted (i.e., subsidized lending, tax preferences, and cross-sectoral subsidies) subsidies are examined, as well
as resource rent and conservation subsidies. The report differenti-
ates between those subsidies that tend to increase fishing effort and
capacity and those that are intended to reduce effort and capacity.

The author argues that subsidies reform be integrated into broader efforts to improve the sustainability of fishery resources.

Sutinen, J. G., A. Rieser, and J. R. Gauvin. 1990. Measuring and Explaining Noncompliance in Federally Managed Fisheries. *Ocean Development and International Law* 21:335–72.

These authors state that—despite its large role in the failure of fishery management programs—understanding of the true nature, extent, and causes of noncompliance is generally inadequate. They briefly review the U.S. federal fishery management and enforcement system and discuss the theory of compliance behavior. On the basis of agency and survey data, they use this theory to describe the extent and patterns of noncompliance in the northeast groundfish fishery. Noncompliance in this fishery is attributed to a combination of poor stock conditions, increased market value, and modest penalties, further aggravated by a negative feedback loop. They recommend targeting chronic violators, concentrating enforcement measures on areas characterized by noncompliance, extending application of management measures to the post-harvesting sectors, substantially increasing the severity of penalties, ensuring input from enforcement authorities early in council deliberations of new fishery management plans, and devoting resources to strengthen the use of noncoercive factors of compliance, such as social pressures, self-interest, inducement, and obligation.

Wilen, J. E. 1979. Fisherman Behavior and the Design of Efficient Fisheries Regulation Programs. *Journal of the Fisheries Research Board of Canada* 36:855–58.

This paper discusses the share-focused behavior of fishermen, whereby input decisions are based on trying to increase or maintain an expected share of total catch revenues. The author explains that regulation of a single component of effort is not effective because share-focused behavior leads to overcapitalization when limits are placed on just one or a few inputs. He describes two options for managers/regulators: (1) to determine and explicitly regulate all key facets of effort, or (2) to design systems that produce an incentive to combine inputs efficiently. He concludes that the second option would effectively remove the motivation for share-focused behavior and produce the incentive to fish efficiently.

Wilson, J. A. 1990. Fishing for Knowledge. *Land Economics* 66(1):1–29.

This article describes how individual incentives for knowledge acquisition conflict with the "rule of capture," which creates a disincentive to share new knowledge necessary for the efficient exploitation of fishery resources. This results in withholding, distortion, and strategic use of information among fishermen. The author discusses the process of fine-grained information sharing among groups/clubs with relatively stable membership, where strong incentives for knowledge acquisition are maintained while information is communicated to a select few and free riders are effectively excluded. He describes each fishery as having unique organizational arrangements developed to maintain incentives for knowledge acquisition. He argues that an assessment of social efficiency must include detailed knowledge related to the efficiency functions of existing institutions and the search/informational/institutional effects of proposed policy changes. He suggests that incentives for the acquisition of new knowledge and information dispersal can be maintained through the creation of institutions and strategic behavior.

Using Scientific Information

Charles, A. T. 1988. Fishery Socioeconomics: A Survey. *Land Economics* 64(3):276–95.

This article reviews and consolidates relevant literature on fishery socioeconomics to examine and discuss the role of multiobjective socioeconomic analysis in fisheries economics and management. The author reviews literature related to the multiple objectives of fishery management (i.e., employment, distributional concerns, and rent generation), income distribution, critiques and examinations of both common and alternative fishery management methods, social and opportunity costs of labor, fishery labor markets, decisionmaking processes of fishermen and fishing communities, data requirements in socioeconomic analysis, and future priorities for socioeconomic research. He suggests two reasons for the difficulties in integrating socioeconomic research and economic analyses: (1) the existence of a "language barrier," and (2) difficulties in choosing fishery management objectives.

Clay, P. M., and E. J. Dolin. 1997. Building Better Social Impact Assessments. *Fisheries* 22(9):12–13.

This article describes the need to better understand the behavior and motivations of fishermen and their interactions with management in order to influence future behavior. The authors explain that regulations creating social and/or economic disruption can lead to noncompliance and that the prediction of regulatory impacts using social impact assessments allows managers to consider alternative tools or increase educational efforts regarding the need for management action. They argue that social impact analyses also assist in formulating balanced, more effective fishery management plans, but that the quality of the assessments needs to be improved by acquiring more comprehensive, relevant, and useful social and economic data and establishing more targeted, well-defined variables and collection procedures. They contend that continued funding is crucial to allow for the creation of social and economic time-series databases and that better social impact analyses will lead to more informed management decisions based on scientific analyses and evaluations instead of anecdotes and emotions.

Fricke, P. 1985. The Use of Sociological Information in the Allocation of Natural Resources by Federal Agencies: A Comparison of Practices. *Rural Sociologist* 5:96–103.

This essay explores differences in the U.S. Forest Service and the National Marine Fisheries Service to explain the variation between the two agencies in the use of sociological information in resource planning and impact analysis. The author reports that, while the U.S. Forest Service ensures the use of social impact assessments and analyses in forest plans, this type of information is rarely included by the National Marine Fisheries Service and the regional councils in fishery management plans. He identifies two factors responsible for the exclusion of this information: uncertainty and organizational climate. These factors are discussed in relation to both agencies/councils and their use/nonuse of interdisciplinary resource management teams.

Garcia, S. M. 1994. The Precautionary Principle: Its Implications in Capture Fisheries Management. *Ocean & Coastal Management* 22:99–125.

This paper describes the precautionary principle and analyzes the scientific, technical, and legal implications of applying the principle to capture fisheries. The principle is examined in relation to conventional management approaches and discussed, primarily, in the international context. Issues such as the use of best scientific evidence and technology, the burden of proof, the role of statistics, assimilative capacity and acceptable levels of impacts, and standards and criteria are addressed. The author suggests elements for the implementation of the precautionary approach in fishery management and argues that this approach can serve to sustain fisheries if it is interpreted and implemented in a reasonable manner.

Lauck, T., C. W. Clark, M. Mangel, and G. R. Munro. Implementing the Precautionary Principle in Fisheries Management through Marine Reserves. *Ecological Applications* 8(1) Supplement:S72–78.

This article discusses the role of marine protected areas in implementing the precautionary principle and sustaining marine fisheries. The authors propose the inclusion of large-scale marine protected areas/marine reserves in marine fishery management strategies to serve as a "bet-hedging component" against uncertainty and management limitations. They outline the desirable features of a marine reserve program and briefly discuss practical issues relating to reserve design. They illustrate how marine protected areas can protect a stock while simultaneously increasing the long-term catch. Finally, they conclude with a list of the important advantages of marine protected areas.

Maiolo, J. R., J. Johnson, and D. Griffith. 1992. Applications of Social Science Theory to Fisheries Management: Three Examples. *Society and Natural Resources* 5:391–407.

This paper describes three cases in which social science research directly assisted fishery resource managers in problem-solving. The first case involves an education and outreach program that was developed for recreational fishermen to promote catch of underutilized species. The second uses social network analysis to identify key opinion leaders and to better understand communication pathways of recreational and commercial fishermen. This exercise resulted in the adoption of a new protocol in the selection of advisers by the South Atlantic Fishery Management Council. The third case analyzes general social and cultural aspects of the menhaden conflict. The authors argue that an understanding of societal aspects is critical to the success of fishery management. They recommend additional funding to promote social science research.

National Research Council. 1998. *Improving Fish Stock Assessments.* Washington, DC: National Academy Press.

This study responds to a request by the National Marine Fisheries Service to conduct a broad review of U.S. stock assessment methods and models. Five different models are evaluated: a production model, a delay-difference model, and three age-structured models. The publication reviews data collection and assessment methods, model performance, use of harvest strategies, peer review of assessments and assessment methods, and education and training of stock assessment scientists. It concludes with recommendations for new approaches.

Orbach, M. 1995. Social Scientific Contributions to Coastal Policy Making. In *Improving Interactions between Coastal Science and Policy: Proceedings of the California Symposium,* 49–59. Washington, DC: National Academy Press.

This article describes social sciences as an essential component of successful environmental policy development and implementation. The author defines two groups that primarily represent the cultural ecology of coastal and marine systems: (1) those who develop the policy that governs the system, and (2) those who are affected by the policy. He argues that social sciences are fundamental to all stages of environmental policy development because the entire process involves value-based decisionmaking. He describes three ways in which social scientists can currently contribute information to the environmental policy management process: (1) as employees of public agencies, (2) as advisers on political advisory boards, and (3) as researchers. He identifies impediments to the use of social science data related to the perceived nature of social (versus natural) sciences, limited support, and the translation of social science data into the policy process.

Sylvia, G. 1992. Concepts in Fisheries Management: Interdisciplinary Gestalts and Socioeconomic Policy Models. *Society and Natural Resources* 5:115–33.

This paper explores new directions in fishery research related to rights-based management systems, fishermen behavior, interdisciplinary research, and multiobjective modeling that can be used to improve the fisheries policy process and increase social benefits. A numerical multiobjective biosocioeconomic policy model is utilized to illustrate the value of interdisciplinary analysis and to analyze the effectiveness of the maximum sustainable yield and optimum yield concepts. The paper concludes with a discussion of the challenges (i.e., data requirements, dynamic systems) faced in developing and implementing comprehensive interdisciplinary analysis and the importance of meeting these challenges.

Vanderpool, C. K. 1987. Sociology Theory and Methods: Social Impact Assessment and Fisheries. *Transactions of the American Fisheries Society* 116: 479–85.

This paper defines social impact assessments and their role in the development of fishery management plans and related policy. The author argues that social impact analyses need to be legitimized and strengthened through clarification of objectives and limitations, an increased store of baseline data, and an assessment of costs and difficulties. He identifies the need for baseline data derived from systematic and comparative studies across different segments of the fishing industry and related social organizations as the central concern of social impact analyses in fishery research. He argues that the historical comparisons derived from these data are essential to capturing the dynamics of stability and change in fisheries systems. He concludes with six additional priorities for sociological research.

Evaluating Fishery Performance

Griffith, D., and C. L. Dyer. 1996. *Appraisal of the Social and Cultural Aspects of the Multispecies Groundfish Fishery in New England and the Mid-Atlantic Regions*. Bethesda, MD: Aguirre International.

This study evaluates the social and economic crises faced by fishing communities in New England and the Mid-Atlantic, as a result of regulatory changes related to the multispecies groundfish fishery. The purpose of the study is to provide managers with baseline social and cultural information that will assist in the development of regulations that minimize negative impact to fishing communities. The report assesses industry demographics and degree of community dependence on the groundfish fishery for five primary ports and nine secondary ports along the northeast coast. In addition, it identifies social science databases that could be used as information sources for follow-up studies, as well as in predicting social impacts of regulations on fishing communities.

Heen, K. 1989. Impact Analysis of Multispecies Marine Resource Management. *Marine Resource Economics* 6:331–48.

This study evaluates the impact of multispecies marine resource management on regional employment and income by combining multispecies bioeconomic modeling and input-output analysis. A regional income-and-employment-impact-maximizing model of marine resource utilization is developed and implemented using data from the Barents Sea fisheries and the input-output table for northern Norway. The results are compared with open access and resource rent maximization solutions.

Iudicello, S. 1996. Overfishing Lures Legislative Reforms. *Forum for Applied Research and Public Policy* (Summer):19–23.

This essay evaluates the MFCMA's success in meeting fishery conservation objectives. The author begins with a review of the current state of U.S. fisheries under the act. She then briefly describes the act, important factors leading to its implementation, and weaknesses that inhibit the achievement of conservation objectives. These include ambiguities, the lack of specific time frames in which to rebuild stocks, non- or misuse of scientific information by the councils, out-of-date environmental impact statements, single-species management, and conflict of interest. She briefly discusses the role of subsidies in the overcapitalization of fishing fleets and highlights potential problems related to limited entry. Finally, she concludes with the Marine Fish Conservation Network's recommendations of amendments to address problems of overfishing, overcapitalization, and conflict of interest.

Marasco, R. J., and M. L. Miller. 1988. The Role of Objectives in Fisheries Management. In *Fishery Science and Management,* edited by W. S. Wooster, 171–83. New York: Springer-Verlag.

This paper describes how fishery management plan objectives are formed as an outcome of the development process rather than defined at the onset. The authors define constraints of the regional councils as limited time and funding, imperfect information, limited rationality, a diversity of preferences, and ill-structured, complex problems. They exemplify this with a review of the North Pacific Fishery Management Council and the groundfish fishery. They argue

that scientists should inform the management process but sympa-
thize with the political constraints of managers/decisionmakers who
must consider input from all stakeholders. Finally, they provide four
recommendations for the successful application of science in fishery
management: (1) scientists should use a multidisciplinary approach,
recognizing that objectives will be driven by social and economic factors; (2) research topics should be addressed by the appropriate dis-
cipline; (3) science should not be driven by advocacy; and (4)
scientists should recognize their role as researchers and not policymakers.

National Marine Fisheries Service. 1999. *The IFQ Program: 1999 Report to the
Fleet*. Juneau, AK: National Marine Fisheries Service, Alaska Region.

In addition to information for the 1999 season, this publication provides a review of the 1998 season in terms of allocations and landings, rate of harvest, location of landings, hired skipper activity,
overages and underages, transaction terminals use, registered buyer
information, enforcement activities, and vessel safety statistics. It
also assesses individual fishing quota program performance in terms
of initial issuance of quota share, determinations, and appeals; quota
share transfer activity; consolidation of quota share; and vessel participation. The last section contains a brief history and description of
the IFQ program and a directory of useful contact information.

National Research Council. 1994. *Improving the Management of U.S. Marine
Fisheries*. Washington, DC: National Academy Press.

This publication reports the results of a study undertaken to assess the
effectiveness of U.S. fishery management. It discusses the MFCMA
and identifies critical issues in U.S. fishery management related to
overfishing, institutional structure, science and data quality, and scientific approach. It outlines policy recommendations for congressional consideration during MFCMA reauthorization.

North Pacific Fishery Management Council. 1994. *Faces of the Fisheries*.
Anchorage, AK: North Pacific Fishery Management Council.

This series of regional profiles, commissioned by the North Pacific
Fishery Management Council in concert with the National Marine
Fisheries Service, is intended to provide a snapshot of various coastal
communities, highlighting their involvement in fisheries off Alaska.
Community profiles were developed for western Alaska, Pribilof
Islands, Alaska Peninsula/Aleutian Islands, south central Alaska,
Prince William Sound, Kodiak Island, southeast Alaska, Washington
(Puget Sound), and Oregon.

Norton, V. J., M. M. Miller, and E. Kenney. 1984. Indexing the Economic Health
of the U.S. Fishing Industry's Harvesting Sector. Presented at the Eighth

Annual Seminar of the Center for Oceans Law and Policy of the University
of Virginia, Cancun, Mexico, January 1984.

This study reports on national trends in fisheries and develops a sim-
plified method of evaluating the economic health of the fishing industry. The Industry Health Index is based on three components: (1)
output price (ex-vessel price); (2) input costs (fuel, labor, repairs,
etc.); and (3) productivity (catch/unit of fishing effort). The index is
used to assess the economic health of eight major U.S. commercial
fisheries, including scallop, Maine lobster, New England otter trawl,
gulf shrimp, menhaden, surf clam, tuna, and king and tanner crab.

Propst, D. B., and D. G. Gavrilis. 1987. Role of Economic Impact Assessment
Procedures in Recreational Fisheries Management. *Transactions of the American Fisheries Society* 116:450–60.

This paper argues the importance of economic impact assessments
in evaluating the effects of policy and investment decisions and discusses ways in which economic impact assessment results can be
most useful to decisionmakers in recreational fishery management.
The authors discuss appropriate uses of multipliers, the linkage
between economic impact assessment and benefit-cost analysis, and
methodological concerns and alternatives. They indicate that,
regardless of the method or model chosen, economic impact assessments in the natural resources field are limited by data quality.

Sissenwine, M. P., and J. E. Kirkley. 1992. Fishery Management Techniques:
Practical Aspects and Limitations. *Marine Policy* 6(1):43–57.

This study provides an evaluation of a wide range of fishery management techniques. The authors begin with a brief discussion of evaluation criteria, followed by a description of the practical aspects and
limitations of active (catch and effort restrictions) versus passive (i.e.,
time/area closures and minimum size restrictions) management. The
benefits of effort versus catch restrictions as means of active regulation
are discussed. They consider the use of incentive/disincentive programs and price control as socioeconomic methods for influencing
fishing effort and examine various techniques that may be used to
allocate catch or benefits, including trip limits, vessel allocations, and
controlled access. Finally, they recommend that additional methods of
allocation be considered in the future.

Looking Ahead

Charles, A. T. 1994. Towards Sustainability: The Fishery Experience. *Ecological Economics* 11:201–11.

This paper presents a historical review of fundamental concepts and
management practices related to ecological, socioeconomic, and
community sustainability. The author synthesizes these concepts in

a framework described as a "sustainability triangle," which is offered as a tool to evaluate ecological, socioeconomic, community, and institutional sustainability. He concludes with an analysis of potential policy directions for sustainable development. These include the use of adaptive management, the development of integrated strategies to cope with the complex interactions between components of the fishery management system, decentralization of control and decisionmaking, the establishment of property rights system, and the combination of comprehensive fishery planning with economic diversification.

Costanza, R., F. Andrade, P. Antunes, M. van den Belt, D. Boersma, D. F. Boesch, F. Catarino, S. Hanna, K. Limburg, B. Low, M. Molitor, J. G. Pereira, S. Rayner, R. Santos, J. Wilson, and M. Young. 1998. Principles for Sustainable Governance of the Oceans. *Science* 281:198–99.

This article identifies the major problems facing the oceans and argues that the key to achieving optimal governance of the oceans lies in adopting management strategies that allow for uncertainty. The authors reason that this adaptive management framework should be structured around six key principles related to responsibility, scale-matching, precaution, adaptive management, full cost allocation, and participation. Finally, they provide examples of institutional strategies that are capable of incorporating many of the principles simultaneously. These include share-based and comanaged fisheries, integrated watershed management, environmental bonding, and marine protected areas.

Iudicello, S., S. Burns, and A. Oliver. 1996. Putting Conservation into the Fishery Conservation and Management Act: The Public Interest in Magnuson Reauthorization. *Tulane Environmental Law Journal* 9:339–47.

This paper reviews the reauthorization of the MFCMA in terms of the public interest. The authors discuss the "common" utilization of fisheries resources, contrasting this with that of other natural resources. They highlight the increased involvement of fishery conservation and environmental groups in the reauthorization debate and describe three conservation amendments designed by a coalition of several of these groups to protect habitat, limit catches, and reduce bycatch and discards. They provide an informed review of the congressional debate on these amendments and end with a brief discussion on overcapitalization and the debate on limited access.

National Research Council. 1998. *Sustaining Marine Fisheries*. Washington, DC: National Academy Press.

This report presents a holistic overview of the problems behind the decline of marine fisheries. It examines issues related to sustainability, the current state of marine fisheries, and the effects of fishing on marine ecosystem structure and function. It discusses options for achieving sustainability—including ecosystem-based approaches to fishery management—and concludes with specific recommendations to significantly reduce fishing mortality. These relate to conservative approaches, ecosystem management, uncertainty, excess capacity, marine protected areas, bycatch and discards, technology, institutional structures, socioeconomic incentives, and information needs.

Organisation for Economic Cooperation and Development. 1997. *Towards Sustainable Fisheries*. Paris: OECD.

This publication begins with a statement adopted by the OECD Committee for Fisheries on the "Study on Economic Aspects of Management of Living Marine Resources" and related to free access regimes and the use of incentives. It provides an assessment of the likely consequences of different fishery management approaches from an economic perspective. Specifically, it examines the historical background of fishery management; highlights fishery management trends since extended jurisdiction; examines fisheries and fishery management in OECD member countries; discusses the use of various management instruments (i.e., output and input controls, technical measures) and their economic, biological, social, and administrative consequences; and looks at the consequences of institutional characteristics.

Sissenwine, M. P., and A. A. Rosenberg. 1993. Marine Fisheries at a Critical Juncture. *Fisheries* 18(10):6–10.

This paper describes problems in fishery management (overutilization, overcapitalization, and resource depletion); identifies the causes behind these problems (open access, uncertain scientific information, and risk-prone fishery management decisions); and evidences ways in which the public, Congress, the National Marine Fisheries Service, the councils, and the industry are working toward the resolution of these problems. The author identifies lost opportunities in U.S. fisheries in terms of net value and long-term potential yield. He reviews national and global production trends and discusses the status of U.S. fisheries, as reported in the 1992 publication *Our Living Oceans: Report on the Status of U.S. Living Marine Resources*. He concludes with a suggestion of two critical management questions to be addressed in the future. Who will pay the cost of correcting problems? And, who will receive the benefits?

Speir, J. (ed.). 1998. *Sustainable Fisheries for the Twenty-first Century? A Critical Examination of Issues Associated with Implementing the Sustainable Fisheries Act*. New Orleans, LA: Tulane University.

This book is a collection of essays presented in the fall of 1997 at the

Tulane Law School in New Orleans. The essays examine critical issues associated with implementing the 1996 MSFCMA provisions related to overfishing, essential fish habitat, bycatch, international management, and the use of individual transferable quotas. It concludes with transcriptions of three ancillary panel discussions recorded at the meeting on the subjects of coral reefs, hypoxic zones, and marsh management.

Waldeck, D. A., and E. H. Buck. 1999. *The Magnuson-Stevens Fishery Conservation and Management Act: Reauthorization Issues for the 106th Congress. CRS Report for Congress*, 31 May 1999, Washington, DC.

This living document reviews reauthorization issues solicited from multisectoral participants in the fishery management system for the 1999/2000 MSFCMA reauthorization debate. The issues described within include individual quota management programs, regional council decisions, bycatch, marine protected areas, essential fish habitat, fishery management data collection, the definition of fishing community, the fishery management plan review process, highly migratory species, fees, and ecosystem health and long-term resource productivity.

Walters, C. J., and C. S. Holling. 1990. Large-Scale Management Experiments and Learning by Doing. *Ecology* 71(6):2060–68.

This article explains that ecosystem complexity causes major resource management and policy changes to essentially act as "perturbation experiments," generating highly uncertain results. The authors discuss passive versus active adaptive management and highlight challenges for designing management programs. These involve embracing/exposing uncertainty, choosing the optimum number of experimental replicates, scale and replication, assessment of transient responses, and prioritization of effort investment.

World Wildlife Fund. 1997. *Subsidies and Depletion of World Fisheries: Case Studies*. Washington, DC: World Wildlife Fund.

This book contains four case studies commissioned by the WWF to promote an informed discussion of future policy options with respect to fisheries subsidies: (1) Euro-African Fishing Agreements: Subsidizing Overfishing in African Waters; (2) The Newfoundland Fishery: Past, Present, and Future; (3) Theory and Practice of Fishing Vessel Buyback Programs (with special section on the New England Buyback Program); and (4) Effects of Japanese Government Subsidies of Distant Water Tuna Fleets. These reports examine the connection between government subsidies and fishing fleet overcapitalization and demonstrate the barriers presented by subsidy programs to achieving sustainable fisheries.

ABOUT THE AUTHORS

Richard Allen has been a commercial fisherman for 33 years and currently fishes for lobsters from the port of Pt. Judith, Rhode Island. Since 1972 he has represented the fishing industry on a variety of public policy issues. Mr. Allen served as a member of the New England Fishery Management Council for 9 years, ending in 1995, and was a commissioner on the Atlantic States Marine Fisheries Commission for 10 years. He is a former member of the National Sea Grant Review Panel, serving one term as its chairman, and is currently a member of the National Marine Fisheries Service's East Coast Advisory Panel on Individual Fishing Quotas. Allen is a graduate of the University of Rhode Island's Fisheries and Marine Technology, Resource Development, and Master of Marine Affairs programs.

Heather Blough has a B.S. in Marine Biology from the University of North Carolina, Wilmington, and is presently pursuing a graduate degree in environmental sciences and policy at Johns Hopkins University. In addition to her work at The Heinz Center, Ms. Blough has served as a researcher and writer on several research projects at the National Academy of Sciences Ocean Studies Board and the Institut de Ciències del Mar in Barcelona, Spain.

Susan Hanna is professor of marine economics at Oregon State University, on leave to The Heinz Center during 1998–99 to direct the Managing U.S. Marine Fisheries program. Dr. Hanna serves on several scientific advisory committees for fisheries and marine resources, including the Science Advisory Board of the National Oceanic and Atmospheric Administration, the Marine Fisheries Advisory Committee of the National Marine Fisheries Service, and the Scientific and Statistical Committee of the Pacific Fishery Management Council. She has also served on the Ocean Studies Board of the National Research Council, the Outer Continental Shelf Scientific Committee of the Department of the Interior, and several National Research Council committees related to marine science and policy, including the Committee to Review Individual Fishing Quotas. Her research areas include the economics of natural resource management, property rights in fisheries, and the economic history of fishing. She is the author of numerous articles and the editor of three books about natural resources and their management.

Suzanne Iudicello is a researcher and writer and operates Junkyard Dogfish Consulting. She is a member of the Marine Fisheries Advisory Committee of the National Marine Fisheries Service. Until April 1998, Ms. Iudicello was special counsel for fisheries for the Center for Marine Conservation, where she served as vice president for programs from 1992 to 1997. Before that, she served as senior program counsel. Iudicello was a leader in the coalition of conservation groups that worked for the changes encompassed in the Sustainable Fisheries Act of 1996. She has written and spoken widely on fishery issues, particularly bycatch. In addition to working at the Center for Marine Conservation for 10 years, Ms. Iudicello represented the Alaska governor's office on Capitol Hill dealing with fisheries and environment issues.

Gary Matlock is the director of the Office of Sustainable Fisheries in the National Marine Fisheries Service. The office provides the service with the principal source of advice and guidance on matters relating to fishery management, international fisheries, fisheries utilization, and aquaculture and administers financial assistance programs for fisheries. Dr. Matlock has also served in the National Marine Fisheries Service as the acting regional director of the Southwest Region and Program Management Officer in Headquarters. Prior to joining the National Marine Fisheries Service, Dr. Matlock spent his fisheries career with the Texas Parks and Wildlife Department, beginning as a fisheries biologist responsible for developing a monitoring program for adult finfish in Texas bays and advancing to director of fisheries. Dr. Matlock has served as a voting member on three regional fishery management councils and has been involved in the development and implementation of fishery management plans and in efforts to reform the federal fishery regulatory process, for which he received a Department of Commerce Gold Medal in 1996. He has published widely on topics of fisheries management and aquaculture, including issues of biology, sociology, and economics.

Bonnie J. McCay is professor of anthropology in the Department of Human Ecology at Rutgers the State University, where she is also associate director for marine and coastal studies of the Ecopolicy Center. Her research areas include the social implications of fishery management, user group participation in regulatory processes, the culture of fishing communities, fishery comanagement, and the socioeconomic consequences of various property rights regimes in fishery resources. Dr. McCay serves on several scientific advisory committees for fisheries and marine resources, including the Committee on Economics and Social Sciences of the Atlantic States Marine Fisheries Commission, and the Scientific and Statistical Committee of the Mid-Atlantic Fishery Management Council. She has also served on several National Research Council committees related to marine science and policy, including the Committee on Ecosystem Management for Sustainable Marine Fisheries and the Committee to Review Individual Fishing Quotas. She has published books and articles on the culture of the commons, participatory management, and property rights in fisheries.

INDEX